WHAT WE
NEED TO
DO NOW

FOR A ZERO CARBON SOCIETY

First published in Great Britain in 2020 by
Profile Books Ltd
29 Cloth Fair, Barbican,
London EC1A 7JQ

www.profilebooks.com

1 3 5 7 9 10 8 6 4 2

Typeset in Bembo and Zephyr
to a design by Henry Iles.

A CIP catalogue record for this book is available from the
British Library.

ISBN 978-1788164771
e-ISBN 978–1782836667

Printed and bound by CPI Group (UK) Ltd, Croydon, CR0 4YY

WHAT WE NEED TO DO NOW

FOR A ZERO CARBON SOCIETY

Chris Goodall

PROFILE BOOKS

'The world is waking up.
And change is coming,
whether you like it or not'

Greta Thunberg at the United Nations
23 September 2019

Contents

NOTES & SOURCES

For sources of data throughout this book, most of
them online, see the 'What we need to do now'
section on my **Carbon Commentary** website
(*www.carboncommentary.com*). The website includes
further analysis and updates on topics featured in
the book, and you can also subscribe
to a free regular newsletter.

INTRODUCTION

A NEW DEAL FOR CLIMATE

We need to create a zero carbon world by 2050 - or earlier

It feels like a turning point. In the past year, tens of national governments have stated that they will cease emitting greenhouse gases by 2050. The UK did so in May 2019, declaring a 'climate emergency'. Scotland made a commitment to cut emissions to zero by 2045, planning to get three quarters of the way there by 2030. Among our European neighbours, France and Spain have also legislated for zero greenhouse gases by 2050. But the big news of the year was China's declaration that it will go carbon zero 'before 2060' and that its emissions will peak by 2030.

Many cities, regions and states have also set challenging targets while a rapidly rising number of major companies have pledged to getting to net zero, often by 2030 or even earlier. ('Net zero' means that any remaining greenhouse gases are at least counterbalanced by extraction of CO_2 from the atmosphere by trees and/or technology.)

Even twelve months ago, these statements of intent seemed very unlikely to be made. But intense public concern about the environment – combined with a growing sense that advanced economies really do have the capacity to make the transition away from fossil fuels – has increased the willingness of companies, cities and countries to make promises on climate change.

In almost all cases, that is exactly what these statements are – promises. They have rarely been accompanied by even the most general plans for how the move away from greenhouse gases will be achieved. The UK government can make a case for being better than most, as in May 2019 its principal advisers, the Committee on Climate Change, issued a report on how it saw the path to zero emissions. But it is an inadequate blueprint, all too heavy on statements of desirability rather than actual plans.

WHY THE UK NEEDS TO ACHIEVE ZERO EMISSIONS

Each tonne of carbon dioxide emitted to the atmosphere adds to the climate problem. The CO_2 will typically remain in the air for hundreds of years, unless it is removed from the atmosphere. Fossil fuel use today will therefore still be causing problems in the lifetimes of our great grandchildren. It is the stock of greenhouse gases in the atmosphere that determines how hot it will get, not how much is emitted in any single year.

Climate researchers often talk of a global 'carbon budget', which is the total amount of CO_2 and other greenhouse gases the world can emit and still remain below global heating of 1.5 or 2 degrees Celsius*. (So far, the we have seen an average temperature increase of about 1 degree over pre-industrial levels). There's considerable debate over these figures but in January 2018 the Intergovernmental Panel on Climate Change (IPCC) said the world should emit not more than 420 gigatonnes of carbon dioxide to have a 67 per cent chance of avoiding a rise of 1.5 degrees. Today that figure is down to less than 350 gigatonnes and global emissions are running at around 40 gigatonnes each year. This means we probably need to achieve zero global emissions by 2030–35 to keep total heating below 1.5 degrees, and 2040–50 for a 2 degree target.

Scientists also tell us that, if we do increase temperatures by more than 2 degrees, we face a much increased risk of unleashing further heating. 'Feedbacks' such as melting ice reducing the reflectivity of the world's surface, thus retaining more heat, become more likely as global heating progresses.

Reducing emissions to zero within thirty years should be the world's most important objective. Even a 2 degree temperature rise will have very substantial effects around the world, including intense rainfall and flooding, rising sea levels and periods of searing drought.

---

* All temperatures in this book are given in degrees Celsius (C).

HOW THE UK PLAYS ITS PART

The purpose of this book is to give an outline of the strategy the UK needs to adopt to address the climate threat – *what we need to do now*. It is a plan that will be similar to that of most other European countries, though each has different challenges and advantages. The UK, for example, has old and poor housing stock, but viable solar power and massive potential for onshore and offshore wind. We also have less than the European average forest cover, and can do much good by a radical reorganisation of our farmland.

We should pursue this strategy, or better alternatives, both because our country should play its part in addressing the global climate threat and because an effective zero carbon strategy will provide multiple benefits to our fellow citizens, both economic and social. Some would argue that the UK has a particular responsibility for leading the way in reducing emissions. We began the Industrial Revolution and our historic wealth is based on fossil fuels. But this is no time for historical debate. We need to act now, in concert with Europe and the world. We can and should become leaders in zero emissions systems and technology.

The UK has made decent progress in recent decades, cutting domestic emissions by 43 per cent since 1990, although a rising volume of imports with high carbon footprints are not included in this figure, nor international aviation and shipping. If all these factors are included, the figures may be closer to 10 per cent (as Greta Thunberg told the UK's MPs in 2019). And it has to be said that most of these reductions have been achieved in relatively easy areas, for example, from the decommissioning of coal-fired power stations. In recent years, progress has slowed, just when it needs to be ramped up to lightning speed. Most independent sources see the UK missing its existing official targets from 2023 onwards.

Meeting the short- term aims, and then cutting our emissions to zero within a few further decades, is a hugely challenging task, requiring action across every part of our economy and society. The incremental, cautious and incomplete programmes the UK government and its advisers are proposing are unlikely to be sufficiently radical. We need a New Deal for Climate. And, as I hope to show in this book, this should be a New Deal with real benefits for those who need it most – giving new purpose to the old industrial towns, renovating our housing stock with effective insulation, getting polluting traffic off the roads to create clean air environments. The big gain is to address climate change. But, in parallel, a Green New Deal offers huge possibilities for improving our quality of life and in creating a fairer society.

A GREEN NEW DEAL

The chapters of this book cover matters as diverse as energy supply, wood cultivation and the fabrics used for our clothes, as well as taxation and research. We need to take action across all of these areas. Addressing climate change isn't just a matter of increasing the percentage of our electricity that comes from renewable sources. It requires coordinated planning across the full range of human economic activities, ranging from what we eat and how we heat our homes through to how we reduce our reliance on cement. Failure to take significant action in any one of these areas means that we will probably fail to reach net zero emissions by our target date.

Of course, all these steps will need the informed support of people across society. In each chapter, I look for ways to ensure that the less prosperous members of society are net beneficiaries of the changes proposed. The original 'New Deal' in President Roosevelt's 1930s America had a similar objective. Without an overwhelming commitment to greater social justice, a democratic society is unlikely to obtain consent for the painful, expensive and complex changes necessary to move us from a society entirely reliant on fossil fuels to one entirely free of them in just thirty years. For example, without a plan to find alternative liveli-hoods for hill-sheep farmers, how do we get support for radical plans for reforestation of most of the UK's uplands? How will we obtain consensus about giving over the land around cities to solar power unless we deliver lower elec-tricity bills for those in energy poverty? And how do we reduce (as we must) the sales of clothing when over a third of a million people work in the fashion business?

Some argue that a transition to zero emissions is impossible in a world controlled by short-term modern capitalism, noting that few shareholder-owned companies have done much to speed up the energy transition (although there are notable exceptions, often from Nordic Europe). Most fossil fuel businesses have doggedly opposed rapid change at the same time as shamelessly and relentlessly advertising minor initiatives to reduce responsibility for climate breakdown. However, my contention in this book is that a global carbon tax at a high enough level could rapidly rotate fossil fuel companies into allies in combating climate change. We desperately need their skills in allocating capital and managing the giant projects necessary to help society implement the massive economic transformations that are needed over the next decades.

I'm encouraged in such optimism by a large number of private conversations I have had over the last few years with employees of fossil fuel companies. I believe that these people are as eager as the rest of us to see progress on climate change. Typically, they know far more about the threat of climate change than the rest of us. They often feel shame at what their employers are doing, and many executives very much want carbon taxation because it will help them escape the financial imperative of continuing to extract and burn oil and gas. A properly designed carbon tax can acquire their active support.

THE NEW DEAL EXAMPLE

Franklin Roosevelt became US President in 1933 and led the country until his death in 1945. Elected at a moment of intense economic depression, he pursued a group of policies that he called the New Deal. Their purposes were varied. Some were intended to provide employment and others to improve business profitability, to create better housing and transport infrastructure and increased availability of electricity.

Although the New Deal did use government spending to kick-start the American economy, one of Roosevelt's aims was to encourage private capital to begin making investments again. Paralysed by the economic crisis of the Great Depression, banks had sharply cut their lending. His administration put in place measures that helped mortgages and loans grow rapidly. The simple view that the New Deal involved little more than pouring government money into the construction of public infrastructure is incomplete. Roosevelt also used incentives to restart the investment activities of private and municipally owned companies across the country. Similarly, governments today can encourage capital to flow at unprecedented levels into the zero carbon economy.

The speed of Roosevelt's New Deal is inspiring. One of its most effective initiatives was the REA (Rural Electrification Administration), set up to encourage local cooperatives to build electricity supply across the farming regions of the US. The REA increased the percentage of rural homes with electricity from 10 to 40 per cent in just five years between 1935 and 1940. With full support from electorates, countries can make truly striking progress.

A group of insightful UK activists and politicians first proposed a Green New Deal in 2008 and I have used many of their ideas in this book. Much more recently, members of the US Congress, and notably Alexandria Ocasio-Cortez, have proposed an outline plan for the US. The ideas are sketchy, but the group proposes to move the US to 100 per cent renewable electricity and switch to zero emission vehicles, among many other measures. As with

Roosevelt's New Deal, the cutting of carbon emissions is intended to go hand in hand with a redistribution of income towards the less privileged and a systematic devolution of economic power away from the big cities.

Is it possible for a UK government to sharply reduce greenhouse gases within a few decades while helping rather than penalising the least privileged? My conclusion is that it may be easier than we might think, and that, despite political divisions, we can reach consensus over good policy. In March 2019 I watched from the edge of a square in central Nice* as the demonstrators quietly dispersed after one of the weekend gatherings of the Gilets Jaunes. Several hundred people were milling about in the spring sun, wearing the yellow jackets that symbolised affiliation to the populist movement seeking to improve life in the less prosperous areas of their country. A second demonstration then began to form in the square, this time from various ecological groups. The Gilets Jaunes are typically regarded as to the right of centre politically while Greens are generally of the left. But, to my surprise, the yellows and greens mixed comfortably, holding polite conversations about objectives and tactics. I overheard discussions about community transport, organic agriculture and the need for local energy generation. It struck me that these two groups were natural allies, each aiming to create a society that restores economic power to the hands of people outside the capital. We need a similar alliance in

---

* London to Nice is an easy andy enjoyable ten-hour train ride, changing in Paris. Bought well in advance, the cost is little more than the air fare.

the UK to push for an equitable carbon transition. There is no inconsistency in aiming for lower emissions at the same time as improving living standards for the less well-off.

WHAT WE NEED TO DO NOW

In summary, the proposals in this book cover ten key areas for action – starting now, with our current level of emissions, which stand at roughly 450 million tonnes of 'domestic' CO_2 emissions (but as much as 875 million tonnes, when you include all consumer activities).*

》 1 Increase renewable electricity generation twenty-fold. This will ensure that the UK has enough electricity both to meet today's needs and to provide for new uses of electric power, such as battery cars. In periods of surplus generation, I propose that we turn spare electricity into hydrogen. We can then use this hydrogen to make power when wind and sun aren't available, to heat homes and to create synthetic fuels that are chemically identical to fossil gas and oil but which have negligible carbon emissions. I also propose that control of the local energy system is handed to municipally owned utilities, who are encouraged to build local generation facilities (much as Roosevelt did). Towns and cities will be allowed to manage the energy networks within their areas, using advanced digital controls that match supply and demand within microseconds.

》 2 Massively improve the insulation of UK houses. Our focus should be on the homes of the less privileged in order to improve living standards and health, using

---

* See the 'Note on numbers' at the end of this chapter.

'deep refurbishment' techniques, with most components built offsite and then transported to the buildings. Our government should make cheap capital available for all types of insulation improvement and mandate that all new building, including factories and offices, be carbon neutral or better (a key failure of legislation to date).

》 3 Electrify the transport system, starting with cars and then moving on to heavy vehicles. Because of the very high carbon footprint of making cars in the first place, prioritise public transport, car-sharing, walking and cycling in order to reduce vehicle ownership. Switch shipping to electricity and to hydrogen made from renewables. We have to assume aviation continues to use liquid hydrocarbons, but we can make its emissions nearly zero carbon by creating fuels synthetically from hydrogen and captured CO_2.

》 4 Move the food system away from meat, due to the impact on emissions of cows and other animals. Shift to forms of agriculture that do not require animal cultivation or artificial fertilisers. Move towards indoor cultivation of plants, to meat substitutes and some organic agriculture.

》 5 Make the fashion business more sustainable. Clothes manufacture is one of the most damaging sources of greenhouse gases and we need to urgently reduce its effects. It is hard to avoid emissions from either cotton or synthetic fabrics, so the best solution is to buy fewer and longer-lasting clothes. Ideally, we should buy items made from cellulose and keep all our clothes for many years.

》 6 Change technologies for production of steel, cement and fertilisers. The key change needed is to use renewable hydrogen as the heat source in manufacturing, or as an ingredient for making ammonia-based fertilisers.

)) **7** Increase the area of woodland, raising its percentage of cover in the UK to typical European levels. This will ensure that forests increase the carbon dioxide they capture through photosynthesis, offsetting those greenhouse gas emissions we find it hard to avoid.

)) **8** Collect carbon dioxide directly from the air and either sequester it safely or use it to make synthetic, very low carbon chemicals, using the hydrogen generated with surplus electricity. This, again, will help counterbalance remaining greenhouse gas emissions.

)) **9** Introduce a meaningful carbon tax, remitting its proceeds to the less well-off, with the principal objective of incentivising the big fossil fuel companies to switch from oil and gas to zero carbon energy. Capitalism can and should be the servant of the energy transition.

)) **10** Research and plan geoengineering techniques. The world will need to have safe, equitable means to artificially hold down global temperatures. Although 'geoengineering' has its risks, we probably have no alternative if we want to keep global temperatures from rising more than 1.5 or 2 degrees celsius. Even the fastest action now looks insufficient to avert dangerous amounts of climate breakdown without measures to block a portion of the sun's energy.

IS THERE POLITICAL SUPPORT?

Awareness of climate change has reached new levels, globally and in the UK. One British opinion poll in July 2019 indicated that 85 per cent of people are concerned about climate change, up sharply from previous surveys. Over half are 'very concerned'. Almost three quarters said that we are already feeling the effects of higher temperatures, and just over half support bringing forward the target of net zero from 2050. A French poll the same month suggested that almost two thirds of the population agreed that the fight against climate change should be the government's top priority, while in Germany, the environment and energy are ranked as the top political problem, with 62 per cent saying that they are very concerned. Across Europe as a whole, the issue is now seen as the most important issue after immigration.

These are only opinion polls. When it comes to backing the difficult political decisions that will need to be made to revitalise the move away from fossil fuels, support may evaporate. But all today's evidence is that emergency action on climate is more possible than ever before. And electoral support will be enhanced if climate action also allows us to push decisions out to local communities and enhances living standards.

The plan outlined in this book shows how a New Deal for Climate might be constructed. It does not provide every detail of how the strategy would work out, but it does indicate a number of routes that we could travel to achieve our aims. It is not an easy or painless strategy. The plan involves huge capital invest-ments and significant changes to the way we live our lives. But, after a period of transition, living standards will be better and the damage being caused to our en-vironment hugely diminished. We might also have far more satisfying working lives.

It would, of course, be good to believe we could make this change to a low carbon society without disruption and expense. But the policy of gradually edging towards lower emissions is not going to work fast enough. My sense is that a majority of British people now share this view and that it may be possible to act with sufficient vigour to avert the worst effects of climate change. And investment in zero carbon has significant rewards. If we invest trillions in off-shore wind farms and other infrastructure, standard economic theory tells us living standards may temporarily drop. But the US New Deal suggests that may not neccessarily be the case. The Defense Plant Corporation channelled about 25 per cent of US national income towards manufacturing investment in 1940, but the average income in the US rose very steeply during the succeeding five years.

The UK has a dismal record in industrial investment (total investment in machinery and buildings was less than 10 per cent of GDP in machinery and buildings in 2016), and this is one reason why our productivity is so low compared to our European neighbours. A sustained push towards development of a post-carbon economy by investing in the transition away from carbon-based fuels will eventually improve incomes, skill levels and economic output.

Much of this investment will need to be managed locally, possibly by the municipal utilities that I believe should control much of the supply and distribution of electricity and other forms of energy. Democratic and local control of the key enterprises driving the transition to net zero will help ensure continued support.

A NOTE ON NUMBERS

UK government estimates of the country's overall 'domestic' greenhouse gas emissions are about 450 million tonnes a year, of which about 360 million tonnes is carbon dioxide and the rest made up of methane, nitrous oxide and gases containing fluorine.

However, these numbers are significantly underestimated in several areas. They exclude international transport (air travel and ships), as well as emissions from changed land use (notably widespread disturbance of British peatlands). These activities add about 75 million tonnes, taking the UK total to approximately 525 million tonnes. That represents about 8 tonnes of CO_2 per year for each of us (roughly the European average).

However, the 'real' figures are still more worrying. Imported goods and services are unrecorded in the UK national accounts (although the government does publish some estimates provided by academic researchers). Much of our food, clothing, steel and fertilisers are therefore not accounted for, because we are almost wholly dependent on imports. These add about 350 million tonnes – or over 5 tonnes a person per year – to the domestic numbers, making 13 tonnes in total per person. There's some reduction from deducting the emissions generated by UK exports, but this only takes the aggregate number down to about 12 tonnes per person. This is a high figure when compared to our neighbours.

In the following chapters I have tried to provide a sense of the importance of each area of the economy by giving an approximate figure for its share of overall emissions. In some cases, such as energy use, I measure emissions against the UK domestic total. In other instances – food for example – I compare the figure against the UK's overall responsibility for greenhouse gas, including imports. This is because much of the UK's food comes from abroad and so is not included in our national emissions.

In the chapters on energy I use two types of number. The first refers to the capacity of a renewable source of electricity. This is usually expressed in terms of gigawatts, or Gw. A very large offshore wind farm might have a capacity of 1.0 Gw, which is its maximum output when the wind is blowing hard. (To give a sense of scale, the typical UK need for electricity over the course of the entire year is about 35 Gw.)

The second measure is for the annual generation or use of electricity (or other fuels), which is expressed in terawatt hours, or TWh. A terawatt hour is equivalent to the power of 1 Gw over a thousand hours. A 1.0 Gw wind farm working at full capacity for the 8,760 hours of the year would generate a total of 8.76 TWh. In normal circumstances, an offshore wind farm will actually typically generate 45–50 per cent of its maximum output over the course of a year as the wind varies in strength, implying an annual production for our 1.0 Gw wind farm of around 4.2 TWh (about 1.5 per cent of UK annual need). A solar farm will produce a lower average output, possibly achieving 10–12 per cent of its maximum capacity during a full year.

We need these two different types of numbers, because the cost of a source of renewable energy is determined largely by the amount of capacity (in gigawatts) installed, while its value arises from its output of electricity – the number of terawatt hours produced.

CHAPTER 1

GREEN ENERGY

Powering (almost) everything with
wind, sun ... and hydrogen

Electricity generation represents about 15 per cent of the UK's domestic emissions (65 million tonnes), a figure that has fallen sharply as coal power stations have been decommissioned. Only twenty years ago, UK greenhouse gas output was twice as much. Other uses of fossil fuels for energy – mainly gas for heating and oil for transport – push the total to around three quarters of the UK's overall emissions; these areas are mostly covered in later chapters.

My proposal for our route to zero carbon emissions is for a twenty-fold expansion of renewable electricity. This will produce far more power than will be needed, even after we have electrified transport and some heating. But it is essential. If we massively overbuild cheap wind and solar farms we will have surplus electricity, almost all the time, and will be able to close all remaining gas power stations.

We will also need to convert as many activities as possible to using electricity, rather than fossil fuels. This means switching to electric cars and using advanced forms of electric heating for well-insulated homes. This will add substantially to electricity demand. However, we should still have huge amounts to spare, which will be available to turn into hydrogen. Hydrogen can be stored to make power when wind and sun aren't available, and then used, with modified domestic boilers, for heating in homes unsuitable for electric heat, and for low carbon substitutes for chemicals and fuels.

Conversion to an energy system based on renewable electricity, plus the generation and use of hydrogen, is a challenging and expensive task. But, as many other countries are rapidly realising, there may well not be any alternative if we want to reach net zero within a few decades. And, as we will see later, renewables offer a cheaper and more easily installed solution than nuclear power, which just a few years ago seemed an essential option for the move to zero carbon. Nuclear power currently contributes around 15 per cent of the country's electricity generation, but several power stations will need to be retired before 2050 and they seem unlikely to be replaced.

THE ENERGY CHALLENGE

We tend to assume electricity generation is the critical determinant of a country's greenhouse gas emissions, but direct use of oil and gas is actually far more important. Electricity provides less than a fifth of our energy needs in the UK. Oil is far more substantial, largely because of its central role in transport. About a third of the total gas consumed is burnt in power stations to make electricity and the remainder mostly goes to provide heat.

Heating homes and other buildings uses more energy than the all the UK's electricity supply. Therefore cutting fossil fuels out of electricity generation is only a small fraction of the challenge we face. We can move some heating to electricity and will certainly convert cars to battery power, but we will still need large amounts of energy for purposes such as heavy industry and aviation.

To give a sense of scale, the table below shows the relative importance of our main sources of energy. (One terawatt hour, or TWh, is a billion kilowatts consumed for an hour.)

ENERGY DISTRIBUTED TO CUSTOMERS IN THE UK	
Oil	800 TWh
Gas	580 TWh
Electricity	300 TWh
Bioenergy	70 TWh
Coal	30 TWh
Total requirement	1,780 TWh

The UK has become increasingly successful at avoiding using fossil fuels to generate electricity. Low carbon sources, including nuclear power, now provide more than 50 per cent of total power over the course of the year. But the use of oil is not declining and gas demand from UK houses shows little sign of substantial change. We need to find ways of using electricity to replace these fossil fuels.

Most specialists working on the financing of power generation have concluded that building wind and solar farms is now the cheapest way of producing extra electricity. That isn't quite the same as saying that renewables are always the least expensive source of power. An old coal or gas power station, fully depreciated in the accounts of its owner, will still sometimes be less expensive, and will often continue to be operated, even if it does little more than break even financially. The new solar or wind farm that might replace it is costly to build and has to return dividends to its owner.

However, we would struggle to find anyone today who doesn't believe that renewables are the logical choice for new low cost power. That's true almost everywhere and doesn't even take into account the climate benefits.

The cheapest electricity in the world now comes from solar photovoltaics (PV) in sunny countries with stable governments. This last part is important, because the price that needs to be charged to cover the costs of constructing a solar farm crucially depends on how much it costs to borrow money. At an interest rate of 8 per cent, solar power costs half as much again as it does at 4 per cent. Borrowers in rich countries with a history of meeting financial obligations, such as the UK, can obtain finance more cheaply.

In some parts of southern Europe, the USA, Latin America and Australia, the price that solar and wind developers offer in competitive auctions is as little as 2 US cents (about 1.6 pence) per kilowatt hour. These numbers are far below the cost of fossil fuel power, even in the US, with its cheap shale gas; just the fuel necessary to make a kilowatt hour of electricity would be costlier. And, to put these numbers in context, these figures compare to almost 15 pence per kilowatt hour paid by homeowners for electricity in the UK, and even more in countries such as Denmark or Germany.

RENEWABLE OPTIONS FOR THE UK

In the UK, onshore wind farms – and probably offshore turbines in future – are the most cost-effective source of power, although even in the UK solar is already cheap enough to beat fossil fuels. (Photovoltaics strongly prefer the cooler temperatures of our cloudy country to the heat of very sunny places, helping to overcome the disadvantages of our latitude.) No other means of generating electricity approaches these technologies, even without applying a carbon tax on fossil fuels.

Private companies are now offering to build offshore wind farms in UK water if they are paid just 4.5 pence for each kilowatt hour of electricity. This is far cheaper than would be possible from constructing another gas power station. Government estimates suggest that gas might cost almost twice this level in 2025. Offshore wind no longer requires any subsidy from bill payers, and on current trends will result in net payments to consumers.

Despite the undeniable cost advantages of wind and solar, the progress towards a fully renewable electricity system is far too slow around the world. This is perhaps because governments and regulators are tending to protect existing suppliers, and partly because it takes time to develop new sources of power.

Perhaps more important is the perceived difficulty of having 100 per cent of electricity provided by wind and solar, as the sources can be intermittent and thus unreliable. Solar power is almost guaranteed in some countries for many months of the year. Wind, however, is less certain. In the UK, wind speeds cannot be precisely predicted even a few hours in advance, and there can be weeks of 'dull lulls' when neither the wind nor the sun perform as we want them to. Many experts therefore dismiss any idea of basing 100 per cent of UK electricity generation around renewables.

Another concern is that the best locations for renewables are sometimes far away from the cities where the power is needed. Building new pylon lines from the source of electricity to where it is consumed is expensive and time-consuming. In some countries, such as Germany, where additions to the electric grid are most needed, resistance to new pylons striding across open countryside is strong and politicians hesitate before approving links. In the UK, similar problems could impede the development of links from many new onshore wind farm locations or from solar farms in places like Cornwall.

The UK government's energy advisers are typical in therefore advocating an electricity system that employs renewables extensively but also has large numbers of gas-fired power stations that stand ready to fire up when wind and sun aren't available and which can be placed on the main electricity links. But how do we get to net zero if gas remains part of the electricity portfolio? The advisers say that the power stations will need to capture the CO_2 arising from the burning of natural gas and then pipe it into permanent storage, probably in old oilfields in the North Sea.

The 2050 scenario identified by the UK government Committee on Climate Change sees only about 60 per cent of all electricity generation from renewables. The remaining gas generation results in over 150 million tonnes of CO_2 emissions, all collected and permanently stored in depleted North Sea oilfields at substantial expense. That figure is about a third of today's UK greenhouse gases of around 450 million tonnes.

I think we should be very sceptical about the financial implications of this plan, partly because we don't yet have a robust cost for capturing CO_2 from flue gases and then piping it offshore. We need a more radical solution. And, perhaps surprisingly, a group of Scottish islands is providing us with an early prototype of how we might run the UK's entire energy system.

THE PIONEERING ISLANDS

Orkney is a group of eighty or so windswept islands off the northern edge of Scotland. About a quarter of these are inhabited, with a population totalling around 20,000 people, many relying on agriculture for their livelihoods. I was told by my one of my interviewees that there are ten sheep and four cows for each person on the islands. However, in the near future, energy will become a more valuable part of the local economy than farm animals.

Very high wind speeds make onshore turbines a natural choice for electricity supply on the islands. In addition, Europe's main testing centre for wave and tidal power prototypes is sited there because of Orkney's fearsome resources of energy in the sea. However, the electricity cables that link the islands to the Scottish mainland have a very limited capacity to accept exports of energy from wind or wave. The consequence is that the ten or so community-owned large wind turbines on the islands frequently have to be disconnected from the local grid, because too much power is being generated. Orkney itself swings between periods when it is short of electricity (and therefore has to import it from the mainland) and when it has far too much.

Energy remains expensive on the islands, even though the resources available should make it very cheap indeed. One problem is that the electricity distribution network was designed to take power from a small number of huge power stations and transmit it to every single building in the country. Orkney sits at the very edge of the UK grid and the islanders pay higher bills to reflect the complexity of getting power to their offshore homes. A second issue is that other sources of energy, such as gas, are either unavailable or very expensive in Orkney. The temporary disconnections that occur when the winds are blowing too hard are a major financial blow to the communities that own the turbines. The money they get paid depends on power actually being put into the network and so 'curtailment', as it is known, reduces local income.

I spoke to one of the members of a group plotting the move towards greater energy independence for Orkney. Adele Lidderdale told me that she thinks that over 50 per cent of the electricity produced on the islands ends up being wasted, and this has impelled residents to start examining alternative uses for the surplus power. Almost fifteen years ago the islands looked at using periods of excess electricity production for making hydrogen, but it wasn't until 2016 that the first project came to fruition.

Why hydrogen? Because it is a very good means of storing energy and can be made relatively easily using electricity. Apply a voltage across two metal electrodes placed in pure water and hydrogen is given off at one terminal and oxygen at the other. You may well have seen this in a school chemistry experiment – the splitting of water into its two constituent gases (electrolysis) will soon become the cheapest way of generating hydrogen.

Hydrogen gas is highly versatile and can provide the energy for transport or heating, or can be converted back to electricity at a time when it is needed. That is why the islanders have decided to push towards an energy system that is based around this light and potentially universally available gas. Another advantage is that the gas has a high energy content but a near-zero carbon footprint. When burnt in air or chemically reacted in a fuel cell, hydrogen absorbs oxygen and just becomes water again.

Adele Lidderdale works at the local council, part of a wide group of Orkney organisations trying to reduce the need to use fossil fuels across the islands and instead run everything on locally produced electricity, converted into hydrogen when it is being stored for later use. She talked me through the main projects, covering regeneration of electricity, provision of heat and the use of hydrogen for transport.

An electrolyser on the island of Eday works in tandem with a large community-owned turbine. When curtailment is threatened, the electrolyser is switched on. It can accept over half the turbine's maximum power, coming into operation within a few seconds. The company that owns the electrolyser pays the turbine's owners a small fee for the power. The hydrogen that comes out of the electrolyser is stored at high pressure and then shipped by ferry to the main island. At Kirkwall harbour the gas is inserted into a fuel cell, a machine that is essentially a reverse electrolyser. Electricity is regenerated from the Eday hydrogen and oxygen in the atmosphere and is then used to run all the on-board power systems when the ferries are in harbour. Adele said most of the hydrogen generated on Eday is used in this way.

This provides, she told me, a good example of how Orkney takes otherwise useless electricity, turns it into hydrogen and then uses it to provide power at a different place and at a different time.

Adele moved on to explain how they use hydrogen for heat, telling me that another island, Shapinsay, which has about 300 inhabitants, was just about to commission a system using hydrogen to heat its primary school. Again, the hydrogen will come from an electrolyser linked to the community wind turbine. It will replace kerosene, an oil that is similar to aviation fuel. Kerosene is expensive and of course produces greenhouse gases when burnt. Adele told me that the Shapinsay hydrogen heating had been difficult and complex to install. Regulations are burdensome, due to concerns about the dangers of explosion. But, as Orkney gets more experienced at using hydrogen, it will become easier to put new heating systems in place. There's no natural gas on the islands, so there's a strong financial incentive to use hydrogen to replace oil for heating.

We went on to discuss how the local council is using hydrogen vans for some of its transport requirements. A hydrogen filling station had opened on the main island and was storing the gas for use in five converted Renault vans. These hybrid vehicles can use either batteries or hydrogen that runs through a fuel cell to make the electricity to power a motor.

Other projects on Orkney include a plan to put a hydrogen-powered ferry into service in 2021. Short-distance sea transport may switch rapidly to hydrogen, because of cost advantages but also dramatically reduced air pollution levels around ports.

BUILDING AN ENTIRE ENERGY ECONOMY BASED ON RENEWABLES

The proposal in this book builds on the ideas employed in the Orkney project. As on the islands, the UK as a whole needs to convert as many energy uses as possible to electricity. That means, for example, all surface transport, including almost all lorries, should be switched to being battery-powered. This tactic is widely agreed.

The second idea is that we should construct huge amounts of new renewables so that, for all but a few days a year, the whole of the UK's electricity demand is covered, including the new requirements from electric cars and some heating. This avoids the need to construct further gas-fired power stations, which would be obliged to capture the CO_2 from the combustion process. (In practice, even the most efficient process is unlikely to hold on to more than 90 per cent of the CO_2 in the flue gas.)

If we cover the daily electricity demand on all but a few days for the rest of the year, we will have substantial surpluses most of the time. What do we do with these? The answer is simple: just as on Orkney, we use the electricity that we don't need for making hydrogen. The scheme proposed here envisages generating enough electricity to cover both our needs for power and for all the other sources of energy our society needs.

The simple hydrogen molecule will perform three main functions in a green economy:

》 We will use hydrogen to make electricity on those rare occasions when supplies are short. If we experience a shortage of wind for a week or so in the middle of winter, even huge batteries won't meet our needs. But we can easily store hydrogen, ready to be converted back to electricity at a moment's notice. (Later in this chapter, we'll note experiments in the Netherlands that will see hydrogen pumped into airtight salt caverns underground that can store prodigious quantities of energy. Similar storage options are available in the UK.)

》 We can replace natural gas with hydrogen for heating homes and other buildings. This new plan will require us to replace central heating boilers and adjust cookers, but hydrogen is a viable replacement for existing fuels. We will also need substantial storage facilities to make this possible.

》 Perhaps most radically, we will use hydrogen to help chemically create the more complex molecules necessary to replace existing fossil fuels. We sometimes call the gases and liquids used across the modern economy 'hydrocarbons'. They are mixtures of hydrogen and carbon and can be made with well-established chemical engineering processes. For those energy uses that cannot easily use electricity, such as aviation, fuels made by low carbon hydrogen will offer us a direct replacement for oil and its derivatives.

This route potentially gives us a 100 per cent low carbon energy system, entirely based on renewable electricity. It is, in my opinion, the only way of completely decarbonising the modern economy while still allowing us to maintain many aspects of our current lifestyle, at least in relation to energy use. Even a couple of years ago, it might have seemed an unconventional plan, but it is one that has rapidly become mainstream. The extremely conservative International Energy Agency (IEA) issued an extensive report in October 2019 about the potential for low cost hydrogen generated from offshore wind around the world. And several countries, including Germany and Australia, are examining similar schemes.

'Renewables + hydrogen' may also be the cheapest way of getting to net zero, and a good way of ensuring that a larger fraction of the money we spend on energy stays within the UK. It is, of course, important that cash is retained by the communities hosting solar and wind farms, to improve local living standards.

HOW MUCH ELECTRICITY DO WE NEED TO COVER ALL OUR ENERGY NEEDS?

As we noted, electricity currently supplies less than 20 per cent of our energy needs – that is around 300 TWh. But how much will we need to cover our other energy needs – for transport, heating and manufacturing – that are currently catered for by fossil fuels?

First, we need to cover the energy needs for surface transport, which is predominantly road traffic. Today, this uses approximately 500 TWh in the form of petrol and diesel. We can replace that using perhaps 140 TWh of electricity, adding approximately 45 per cent to total generation. The reduction arises because electric vehicles are almost four times as energy-efficient as petrol cars.

Second, we can use electricity for some heating needs. Elsewhere in the book, I express reservations about heat pumps, a technology for heating houses efficiently. But there's no doubt that some homes can be effectively heated with these machines, and as we improve insulation that percentage will rise. I assume that currently 25 per cent of domestic heating can be switched to electricity and this increases demand by about 100 TWh. This is concentrated in winter, of course.

We also need to convert the other uses of coal – chiefly for making steel – to hydrogen. This is dealt with more extensively in Chapter 7. The hydrogen needed for the process is made from electricity via water electrolysis.

Very roughly, these conversions double the current need for electricity to around 550 TWh a year.

However, we will still have substantial amounts of fossil fuels still in use unless they are replaced by low carbon alternatives. These include gas for heating, and oil for aviation and shipping, with a total energy need of almost 600 TWh a year. My proposal that we make these fuels synthetically from hydrogen, or use hydrogen itself, require a further 800 TWh of electricity for electrolysis.

In total, therefore, we will need about 1450 TWh for a fully electric system. At the moment, we generate about 80 TWh from wind, solar and other renewables, excluding biomass. So, to get to a 100 per cent renewable energy system we need to multiply our generation of zero carbon electricity by about twenty times. This sounds ambitious, and it is.

INCREASING CAPACITY TWENTY-FOLD

Let's look at what an increase of twenty times means in terms of the current installed capacity of offshore and onshore wind and solar PV. At the time of writing (late 2019), the UK has about 23 GW of wind (of which 9 GW is offshore) and 13 GW of solar PV. In addition, we have some renewable generating capacity using biomass (mostly wood pellets from North America), a number of hydroelectric dams and some power from anaerobic digesters that burn gas from the rotting of farm waste. Innovators have also installed a tiny number of tidal turbines. I don't think these sources can be expanded significantly, although I would love to be proved wrong, particularly by those collecting energy from the tidal flows of water around the UK or building lagoons to capture the energy in the rise and fall of the tides.

If we multiplied the generating capacity of wind and solar sufficiently to generate 1,450 TWh a year, we would need about 460 GW of wind and 260 GW of solar. Actually, because offshore turbines are very much better at capturing energy than all but the best onshore locations, we might manage with 300 GW offshore and 50 GW onshore.

Increase the output from modern renewables by a factor of twenty and we have solved one of the principal problems of using wind and solar: their intermittency (and unreliability). We have so much capacity to generate electricity that even during 'dull lulls' we will usually have enough power. We will still need batteries to store power over the course of twenty-four hours, because solar energy is only available during the day. The best way of doing this may be to use the storage capacity of car batteries (more about this in Chapter 4) rather than larger commercial storage units.

So, during the vast majority of the year – even when the wind isn't blowing hard or the sun shining – a huge expansion in renewables enables us to meet our daily needs, particularly if batteries match supply and demand over the course of twenty-four hours. We will be generating enough both to meet our current electricity requirements but also to fill up the batteries of a future fully electrified car and lorry fleet and provide the electricity to power some heat pumps in homes and offices.

On most days there will be a huge surplus for making hydrogen. Conversion of electricity using electrolysis will involve some conversion losses, of perhaps around 20 per cent, and this needs to be factored into our calculations.

Is it a viable scheme? There are three key issues. Will the wind and solar generate enough electricity on most days each year,? Is the size of the electricity supply also sufficient to generate enough hydrogen? And how much land and sea area do we need to meet our requirements?

WILL THE WIND AND SOLAR GENERATE ENOUGH ELECTRICITY – AND HYDROGEN?

To answer this question, I looked at the pattern of electricity needs over a full year from 16 July 2018 to 16 July 2019, charting how much power was being used every five minutes, and then comparing it to the amount of renewable energy being produced. I had to make some crude assumptions, but the data seem clear: if the UK had twenty times as much wind and solar capacity as it does at the moment, we would have covered our electricity needs on all but thirty or so days.

I calculated that the average day has a surplus of around 2 TWh hours of electricity that wouldn't be immediately needed. Compare this to a typical day's total electricity demand of about 0.8 TWh across the UK at the moment. The surpluses tends to be higher in winter and lower in summer, because high winds are concentrated from September to March.

The days in which power was insufficient over the day were mostly in December and January during periods when wind speeds were unusually low. The most difficult stretch of time would have been the first days of the 2019 New Year. My simple model showed that the accumulated deficit over the first five days of January was almost 4 TWh. But even this intermittency would have been easy to cope with, because the last few days of December had seen typical daily surpluses of over 2.5 TWh. If we use hydrogen for making electricity in periods of deficit, we would lose some of the energy in the conversion. But, in the worst period of 2018–19, if we had stored the hydrogen from the two days before the start of a period of winter calm, it would have been enough to last nearly a week of low winds when converted back into electricity.

The next question is whether the surplus electricity we obtain over the course of the year is sufficient to cover the non-electricity energy needs of the UK. The figures suggest that we would have about 800 TWh left over, which is enough to cover all our remaining needs for gas for heating and for industrial use, as well substituting for the demand for oil after we have switched all surface vehicles. My central assumption is that any conversion of hydrogen into other chemicals delivers about 50 per cent efficiency. It may well be that this is too pessimistic.

HOW MUCH LAND AND SEA IS NEEDED?

We need to think about the surface area of land or sea required to site the photovoltaic panels and wind turbines. As you would expect, numbers are large. Very roughly, one megawatt (1 MW) of PV solar panels requires 20,000 square metres (2 hectares), so 260 GW will need about 5,000 square kilometres. This can be either on the roofs of buildings or on open countryside. PV will thus occupy about 2 per cent of the UK's 240,000 square kilometres – an area about one third the size of Yorkshire. We should note, too, that the best sites for solar power sit south of a line running roughly from the Wash to the Bristol Channel. Panels put in this area generate more power than those in less sunny regions. Nevertheless we will want to spread the panels out across the UK so that they can be developed and owned by local communities (more on which later).

Our assumption is that we need 300 GW of offshore and 50 GW of onshore to deliver twenty times as much electricity from wind. We need an offshore region of about 250 square kilometres to meet the needs for spacing between turbines. This is a large area, without doubt, but only about three times the size of Dogger Bank, one of the most attractive existing North Sea areas.

We may struggle to find enough excellent locations for offshore wind with relatively shallow water and easy access to the UK coast, but the space seems to be there. I'll refer later to a study that Shell sponsored which suggests the North Sea could offer a total of 900 GW for its neighbouring countries.

Putting 50 GW onshore will multiply the existing land-based capacity installed in the UK about four-fold. Fitting in the extra 37 GW won't be simple, will be strongly opposed in some places, and will require new pylon lines, but will leave the UK with slightly less onshore wind than Germany has today. It's an ambitious target, but within our capability to install.

IS THIS REALLY THE BEST ROUTE TO A LOW CARBON ENERGY SYSTEM?

Even a few years ago, this scheme would have seemed absurd. Renewables were far more expensive than power from gas and coal, which meant that creating hydrogen from electricity was uncompetitive and the idea of using the gas for making synthetic fuels was hardly considered.

This has changed, because renewables are the cheapest way of generating electricity, and almost everyone expects renewable electricity to continue to fall in cost. Forward-looking energy companies across Europe, including Statkraft, the European leader in hydroelectric power, are also forecasting rapid growth in the use of hydrogen as a store for surplus renewable energy. Overbuilding of renewables, complemented by hydrogen from electrolysis, is now seen by some energy commentators as the most likely route to a low carbon energy system.

A few years ago, we might have thought that new nuclear generators might fill the role of renewables today. But the experience around the world of building new power stations has been almost uniformly disastrous. Delays of ten years are common and costs have escalated to entirely unpredicted levels. The generating plant in construction at Flamanville in Normandy is now expected to come online in 2023, eleven years late and at a cost of almost four times the projections when construction began.

Although there is little reason to close the existing nuclear plants in the UK's ageing fleet – not least because they still provide 15 per cent or so of the UK's electricity – the financial argument for backing new development is weak. At today's expected price levels, nuclear power would be at least twice the cost of offshore wind or solar. Nuclear power stations operate continuously, which is why some people prefer them to intermittent renewables. However, if we can use hydrogen as a long-term storage medium when the wind isn't blowing, this advantage disappears.

In fact, the supposed advantage of nuclear power – the delivery of a continuous stream of electricity – is increasingly seen as a disadvantage. Nuclear is inflexible and cannot respond to changes in demand. As a consequence of this, the US government recently started investing in pilot schemes at large nuclear plants that will use electrolysis as a way of productively disposing of surplus power from the power stations when demand is low.

HOW MUCH WOULD THE NEW RENEWABLES STRATEGY COST THE UK?

What would 260 GW of solar PV and 350 GW of wind, mostly offshore, cost? At today's prices the answer is probably about £800bn. Spread over twenty years, that is about 2 per cent of the UK's GDP, or £40bn a year (this is probably an overestimate, as we see further cost declines in renewable technologies).

This is a sizeable investment, but one that will also deliver returns. For at the end of the process of installing this generating capacity, we'd no longer need to buy any natural gas or oil and would have cheap, clean sources for all our energy. Today, the UK spends about £45bn a year on buying fossil fuels, which is slightly more than the investment requirements for 100 per cent renewables. In twenty years, Britain's bills for oil and gas will entirely disappear. We'd be saving enough to fund about a quarter of the National Health Service.

The move to 'renewables plus hydrogen' is even more beneficial because, although the North Sea currently gives the UK about half the oil and gas it burns, our self-sufficiency will fall over the next decades. Massive overbuilding of renewables would offer full independence from the fossil fuels of other countries, whether Russian gas or Iranian oil.

Intriguingly, this is not an original idea. Researching this book, I came across a passage written by the great British biologist J.B.S. Haldane, who foresaw the importance of renewable electricity combined with hydrogen as the basis of the entire energy system as early as 1923:

Ultimately we shall have to tap those intermittent but inexhaustible sources of power, the wind and the sunlight. The problem is simply one of storing their energy in a form as convenient as coal or petrol ... The country will be covered with rows of metallic windmills working electric motors which in their turn supply current at a very high voltage to great electric mains. At suitable distances, there will be great power stations where during windy weather the surplus power will be used for the electrolytic decomposition of water into oxygen and hydrogen. These gasses will be liquified, and stored in vast vacuum jacketed reservoirs, probably sunk in the ground ... In times of calm, the gases will be recombined in explosion motors working dynamos which produce electrical energy once more, or more probably in oxidation cells ... These huge reservoirs of liquified gasses [sic] will enable wind energy to be stored, so that it can be expended for industry, transportation, heating and lighting, as desired. The initial costs will be very considerable, but the running expenses less than those of our present system. Among its more obvious advantages will be the fact that energy will be as cheap in one part of the country as another, so that industry will be greatly decentralized; and that no smoke or ash will be produced.

This is all pretty much what I am proposing in this book. Haldane thought that it might take four hundred years to 'solve the power question in England'. We now know we need to carry it out in the next two decades.

However, we still need to answer a few further key question. Will renewables continue to get cheaper, both here and around the world? Are other countries considering hydrogen as a way of storing surplus energy – and can we learn from them? And how can we use the hydrogen to make low carbon substitutes for the gas and oil we will still need?

WILL SOLAR AND WIND GET CHEAPER?

It is only a slight exaggeration to say that every single forecast for the evolution of the costs of solar and wind power has been proved too pessimistic. The idea that 2019 auctions held by governments and electricity companies around the world could produce bids as low as 2 US cents a kilowatt hour would have seemed foolish even a few years ago. My own book on solar power, *The Switch*, published in 2016, has estimates from analysts at Citibank suggesting that this level wouldn't even be achievable by 2030.

The cost of wind power is also falling unexpectedly fast. In this case, the decline is concentrated in offshore developments. All across Europe, bids are being made by experienced private companies that undercut all other sources of electricity. The size and efficiency of turbines is increasing every year.

But will these falls continue? I see no reason why not. In the case of solar, we will see panels becoming cheaper because of continued sharp improvements in efficiency and changes in materials. Oxford PV, a spin-out from the university, has developed an inexpensive coating on top of a conventional silicon PV cell that will improve the yield by about 25 per cent. Although this technology has been long in gestation, the company says it will be producing hundreds of megawatts of advanced solar panels from its factory in northern Germany by late 2020.

Also in Germany, Heliatek is producing an entirely new form of thin PV material which is printed onto a flexible plastic backing sheet. This extremely light film can be easily pinned to walls and roofs and doesn't require expensive installation. It could go on any surface with a good view of the sun. It isn't as efficient at collecting light as an Oxford PV panel but it will become very cheap.

Technological developments like these will help continue to push the price of solar power down to unprecedented levels. People in the industry now talk of a price as low as 1 US cent per kilowatt hour for solar generated in the best locations. Even in the UK, 3 pence per kilowatt hour looks a feasible target. And the good thing about solar is that it can operate successfully at any scale, whether on the roof of a house or a warehouse, on the side of an office building or across one square kilometre of unproductive farmland. It is the perfect technology for helping restore local control over energy supply.

'ENERGY ISLANDS' FOR OFFSHORE WIND

What about wind power? Will it continue to fall in price as well? The high average wind speeds around the coast of the British Isles means that this question is probably more important to us in the UK than the price of solar.

Bidders for opportunities in the North Sea are already offering prices lower than any fossil fuel generators can match. Offshore wind power developers are relying on increasing turbine sizes and taller towers. This means more electricity generation from the faster wind speeds high in the air. Installing larger turbines in deeper waters adds to costs, but the improvements in electricity yield more than make up for this.

In the next chapter I argue for decentralised control of energy supplies, giving control to municipally owned utilities so that they can provide cheaper, clean power to their citizens, as well as sustaining jobs and local purchasing power. But at the same time we need massive investment in offshore wind farms if we are to completely replace all our oil and gas with electricity and synthetic fuels made from hydrogen and carbon dioxide.

I therefore suggest a very different route for offshore, and it seems already to be taking shape. A group of Dutch companies, led by Tennet, the Dutch equivalent of the UK's National Grid, are at the forefront of a campaign to develop European collaboration in the North Sea to save money and improve energy connections between countries around the basin. This will help further reduce the cost of offshore wind power and also reduce the problems of intermittent supply.

The central idea is that a series of large man-made islands across the shallow parts of the North Sea will be constructed to collect electricity from huge wind farms. Tennet calls these islands 'wind power hubs'. Power generated by local offshore wind farms can be routed to the nearest of these artificial structures and then on to any of the surrounding six or seven countries. At present, a country wanting to build an offshore farm in the North

Sea either builds the cable infrastructure that takes the power from the turbines or demands that the developer installs the link to its home coastline. A new UK wind farm, for example, has to build a connector that takes its power onshore, where it is fed into a substation. The Tennet scheme allows each farm to connect to a hub (of which there might be several in the North Sea), which in turn link to several different countries. The spokes that fan out to the UK, Norway, Denmark, Germany and the Netherlands from the new islands will thus be able to function as indirect connectors between the countries. Power from a wind farm can go to one of the hubs and then on to any of the North Sea countries.

The partners in the consortium stress that this project saves money compared to national development of North Sea wind resources. It allows the energy to be exploited more quickly than any single country could manage. Furthermore, it creates a viable initial plan for the installation of at least 180 GW of power generation and distribution capacity (around five times the UK's current average electricity demand).

The Tennet plan also involves the development of a hydrogen infrastructure so that the power in the wind can be stored when electricity demand is low or gales are blowing across Northern Europe. It recognises that the way to deal with intermittent supply is to build very high renewables capacity and store the surplus as hydrogen.

Another reason for believing in the Tennet idea is that it will create a huge series of projects demanding many hundreds of billions of capital expenditure. An advantage of the project's scale is that the large fossil fuel companies will be interested in participating. Businesses such as Shell and BP currently each invest in the order of $20bn a year in fossil fuel development. They have shied away from renewables because of what they see as the very small scale of most developments in wind and solar. The North Sea is different to any other set of projects and could enable these companies to switch sides towards the clean energy economy. Give Shell a $10bn project, I say, and you will find no company in the world better at completing it on time and on budget.

In a study published in December 2018, Shell and other companies argued for an aggressive target of 900 GW of offshore wind in the North Sea. (This target is consistent with the ambition I set earlier for 300 GW for the UK, which controls much of the area.) Additionally, the businesses behind this report asked for detailed plans for the conversion of temporary surpluses to hydrogen. If we can get the oil and gas majors firmly behind a scheme for renewables plus hydrogen storage, we can make the transition quicker and more cheaply.

100 PER CENT RENEWABLES COMBINED WITH HYDROGEN GENERATION

As we've seen, plans for combining renewable energy with hydrogen production are in development on a huge scale in the North Sea and as a community-based initiative on Orkney. And these are not isolated examples: at various places around the world, projects are being assembled that will use a combination of solar or wind and hydrogen electrolysis to cover energy needs.

Possibly the most important is on the north-west edge of Australia in an area called Pilbara. There, one of the most successful renewables companies in the world is planning to build a 15 GW wind and solar project spread over 6,500 square kilometres. This single project will produce the equivalent of one sixth of the UK's current electricity need. The scheme won't begin construction for several years, but backing is in place from Macquarie, a leading lender to renewables projects, and Vestas, the large Danish wind turbine manufacturer. It has active support from Australian authorities and the traditional owners of the land, the Nyangumarta people. It offers a major new source of income and employment.

Pilbara is unusual in having exceptional wind and solar resources in the same place. The sun shines by day and the wind complements it by tending to be stronger at night. This means that the electrolysers that convert electricity will usually be working twenty-four hours a day, reducing the cost of the hydrogen. Pilbara also has well-developed ports that service the iron ore exports from the region, meaning that the project's output can be shipped easily.

The project developers envisage that about one fifth of the power from the turbines and panels will be taken by large mining companies in the area. The rest will be turned into hydrogen and shipped to Asian countries, possibly in the form of ammonia. This is another simple molecule, containing just hydrogen and nitrogen, which can be transported in bulk easier than pure hydrogen.

A key market will be Japan, which has a shortage of domestic sources of energy and has decided to build its economy around hydrogen. It will put the gas at the centre of the 2020 Olympics, using it for transport but also fuelling the Olympic flame with it. And shipping hydrogen long distances is not expensive. Modern ships are cheap and efficient, and the cost of importing hydrogen will be no more burdensome than oil or gas. Australia is excited by the prospects for building a large export industry from hydrogen, perhaps to replace the current emphasis on shipping coal to power stations across Asia.

The total construction cost of the project will be about £16bn, with very low running costs. This is less than the construction cost of the UK's new nuclear power station at Hinkley Point, which will produce less than half its electricity output. Indeed, if the scheme were in the UK, the electricity produced would be worth almost £3bn at wholesale prices over the last year, meaning payback would be achieved in as little as six years, even without taking into account the extra value of the hydrogen.

The reason I offer these simple calculations is that I want to show how it may already be financially viable to use renewables to provide all types of energy, not just electricity. But the extraordinary scale of this project should also be noted. The Pilbara scheme will occupy an area of 80 by 80 kilometres, demonstrating just how much land will need to be given over to renewables capacity in a low carbon world. A scheme the size of Pilbara is far easier to construct in a very dry and thinly populated region of Australia than it would be in the

UK. However, the shallow waters of the North Sea can provide us with the space for many offshore wind projects of a similar size and potential.

Although nothing proposed in Europe matches the huge scale of the Pilbara scheme, Norway is also pitching to supply Japan with prodigious quantities of hydrogen produced from renewables. And there are many smaller experiments starting across Europe that mirror the Orkney trials. In June 2019 Gasunie, the Dutch gas grid operator, opened the company's first electrolyser linked to renewables. The project will take the output of 1 MW of solar panels, complemented by grid electricity when the sun isn't shining – and store the hydrogen that is produced by the electrolyser for later use.

The aim is to demonstrate how the electricity from intermittent renewables can be stored for long periods and in huge volume. The project is located close to existing salt caverns that store natural gas. If switched from storing natural gas to hydrogen, a single cavern in the northern Netherlands can hold enough energy to provide one day's electricity use for the entire country.

LOW CARBON HYDROGEN AVIATION FUEL

The final building block in the move to a 100 per cent renewable energy system is the use of hydrogen to make gases or liquids that can freely substitute for fossil fuels. Planes, for example, will not switch to battery power or hydrogen within the next few decades and will continue to need liquid hydrocarbons.

Gasunie, the Dutch gas grid operator noted above, is embarking on a project to make hydrogen that is then converted into low carbon aviation fuel. This synthetic liquid is made by chemically combining streams of waste materials, including used cooking oil, with the hydrogen. Tennet and its renewable fuels partner SkyNRG promise to make 100,000 tonnes of fuel a year from 2022 onwards. This is less than 0.1 per cent of the fuel needed for world aviation, but it shows that the idea of using hydrogen to create aviation fuel that can replace fossil fuels is technically feasible. Gasunie is one of the largest natural gas distributors in Europe and seems to be convinced of the viability of the hydrogen route. It asserts that 'we can have the infrastructure in place for the large scale use of hydrogen as a replacement for fossil fuel in place as early as 2030'. It is working on nineteen separate projects around the Netherlands to use it as a replacement for fossil fuels.

Many other projects across Europe are trialling the use of hydrogen as a large-scale storage medium for electricity that would otherwise be wasted. These experiments benefit from the low wholesale price of electricity at times of abundant wind or solar power. As renewables grow across the world, countries face increasingly long periods, sometimes lasting weeks, when electricity supply exceeds demand. This is as true in Britain as elsewhere, particularly when the wind is blowing hard in autumn and spring.

Some of the most promising technologies for using hydrogen to make fuels that are similar to petrol or diesel come from Germany. The country's increasing problems with dealing with large amounts of solar and wind have pushed companies, and the government, towards solutions that use hydrogen, either on its own or in another chemical compound that can store energy.

One such company is Sunfire, a leader in making low carbon replacements for motor fuels and the raw materials for plastics. It is now leading the work to develop an aviation fuel refinery close to Rotterdam airport. This process will have four stages. First, CO_2 is captured directly from the air using equipment provided by a Swiss company called Climeworks (more on this 'direct air capture' technology in Chapter 11), and hydrogen is generated using electrolysis. The two gases are then converted into 'syngas' – a mixture of carbon monoxide and hydrogen – using the 'water gas shift' (WGS) reaction. And finally, using the well-established Fischer Tropsch process, the syngas is refined into a near identical substitute for aviation fuel (with the advantage that almost no pollution from sulphur compounds will be produced when the liquid is combusted). When the fuel is burnt, the CO_2 that was captured by the Climeworks plant will return to the atmosphere. The whole cycle is therefore close to carbon neutral.

Creating low carbon substitutes for oil or gas sounds challenging. But there is nothing innovative about the Fischer Tropsch process (invented in 1925) or making the 'syngas' through the WGS reaction (discovered in 1780).

The first Sunfire plant will produce limited quantities of fuel for flights leaving Rotterdam, and we are still some way from full-scale commercial development, but some variant of the process may become the global standard for producing liquid fuels for aviation and other markets that cannot be converted to electricity.

As with all these technologies, the critical question is not whether the technologies work but rather how fast the cost of renewable electricity falls. This is what principally determines the cost of zero carbon hydrogen. If the hydrogen price falls, synthetic low carbon fuels will become cost-competitive with conventional fossil oil and gas. And, of course, the faster we deploy more renewables, the lower their cost. Driving the pace of wind and solar development will force the cost of low carbon fuels down to parity with oil and gas.

What could the UK do to speed up the growth of synthetic fuels? Amazingly, most of our national research and development is devoted to improving the prospects for nuclear energy and, to a lesser extent, carbon capture and storage, rather than renewable electricity or energy storage. The government has invested minuscule sums into the development of low carbon synthetic fuels, despite the obvious importance of finding replacements for oil and gas. This is an urgent priority and would be an obvious use for the proceeds of any tax on carbon.

CHAPTER 2

LOCAL GRIDS

Taking back local control of our
energy generation and distribution

The UK could make the energy transition to zero carbon cheaper and quicker by introducing local control over energy networks. In this, we would do well to copy the German system, which uses municipally owned companies to operate many electricity and gas systems, as well as to provide services such as broadband and transport. This helps to hold down costs, and increases the ability to manage supply and demand for energy using advanced digital technologies.

Local utilities, many publicly owned, distribute over 60 per cent of all Germany electricity – about a third of which is generated at power plants and renewables sites they themselves own. These *Stadtwerke* – around 700 of them, spread across Germany – control 800,000 kilometres of electricity network, employ 60,000 people and sell €50bn Euros of power to their customers.

This situation couldn't be more different in the UK, where the small group of private distribution companies, many owned by private equity funds domiciled abroad, act to maximise profit rather than the public good.

THE GERMAN MODEL

The largest and best known of the German *stadtwerke* covers the city of Munich. It serves three quarters of a million customers and supplies almost all homes within the city. It owns the local distribution network and large electricity generating plants at hydroelectric dams in the foothills of the Alps. It promises that by 2025 it will be generating as much zero carbon electricity as it currently supplies to its customers over the course of the year. Much of this will not come from local generation but from wind turbines in central Norway. In addition to decarbonising electricity, Stadtwerke München is pushing forward with plans for providing heat supplies without any fossil fuels. It intends to use geothermal reservoirs under the city to provide hot water for district heating systems. The utility already supplies low carbon heat from a district heating system powered by burning wood.

Stadtwerke München also operates the city's telecommunications network, which takes fibre broadband to the door of most of the city's flats and houses, as well as public transport, electric vehicle charging and some housing. The profits of this company, several hundred million euros a year, all go to the city of Munich. It provides jobs for over 9,000 local people.

POWER TO THE PEOPLE

In the UK, to ensure that the energy transition improves the economic circumstances of people in less prosperous parts of the country, we will need to push for three crucial changes.

>> Towns and cities should own and operate their own renewable energy resources as far as possible.

>> Distribution of electricity (and possibly hydrogen for heating) should also be controlled by local bodies.

>> We need to encourage development of urban 'microgrids' that use digital technology to balance supply and demand within a city or region. Trading of electricity directly between producers and users should become the norm.

Essentially, what this plan aims to do is to make it possible for electricity to be generated, distributed, stored, transformed if necessary, and then used, all within the same area of the country.

This isn't as new as it might sound, even in Britain. Down the road from my home in Oxford is a building that once supplied all the electricity for the city. Built in 1892, this coal-fired power station took coal from barges moored on the Thames and used the river for its cooling water. Some 7,000 electric lamps were connected to the innovative local grid within twelve months of the power station opening, and within a few years power, principally for lighting, was distributed over a wide area, including all the university colleges. Though all the fuel came from elsewhere, the power station provided employment and income to Oxford until its closure in 1969. (Incidentally, the developer of this early power station, Thomas Parker, also designed some of the first electric cars.)

The national grid, the UK's countrywide electricity transmission system, was not completed until the mid-1930s. Before then, each power station operated independently or in a local group. My proposal is that we partly replicate the way the electricity supply system worked before the development of the national grid. We may decide to go back to the days of Oxford power station in its early years. The larger national network of power supply will still exist. But, instead of a small number of very large power stations linked by a complex and ever more expensive national and local grids, we should devolve control of electricity supply to a hundred or a thousand urban areas, as in Germany.

In this plan, each town or city could be supplied by a group of nearby solar farms and wind turbines, complemented perhaps by small hydroelectric power schemes on rivers, electricity from anaerobic digestion plants on farms, or hot water from deep geothermal wells. A business owned by each city or region would own and operate the energy generation assets.

GOING LOCAL IN THE UK

Successful attempts have already been made in the UK to establish locally owned electricity and gas companies of this type. Probably the best known is Robin Hood Energy, which is based in Nottingham but also supplies other small utilities around the UK. It has a very good track record; however, it doesn't remotely compare to the largest German municipal utilities. My proposal is that locally owned utilities such as Robin Hood Energy should be encouraged to invest in their generating assets and operate their own electricity, gas and heat networks. The impact on the city of Nottingham from developing its own fully integrated utility would be enormous.

How could Nottingham increase the percentage of its own energy that it sells? The obvious route would be to install solar farms around the city and possibly supplement this by buying the output from wind turbines in much more windy locations, perhaps on the coast of Lincolnshire, fifty miles away.

How feasible is this plan? Let's take a look at what Nottingham would need to meet its electricity demand over the course of the year just from solar. Nottingham would need to build solar farms or solar roofs of about 1.7 GW maximum generating capacity, equivalent to around an eighth of the UK's current solar plants. This would cost just over £1bn today (the figure is likely to decline as the cost of solar power continues to fall). This is equivalent to about £3,000 per Nottingham resident. Local authorities can borrow money at very attractive rates, making it possible to comfortably service the debt and still provide electricity at competitive prices.

The total space needed to meet the city's need for energy would be about 34 square kilometres. That's an area of about six by six kilometres, which is just under half the size of the city of Nottingham. The most obvious local sites for large solar farms are the area's old coal-fired power stations. Several are closed, but their sites have not been redeveloped some fifteen years after they stopped generating. The three still open in 2019 will also have to close within a few years, making space for renewable energy. The biggest, on a site at Ratcliffe on Soar, is less than 15 kilometres from Nottingham centre and on its own offers about a tenth of the space needed. An advantage of using old power stations is that they are already connected to the high-voltage electricity network.

It is local people who should decide where the solar farms go. However, redundant industrial sites, such as derelict power stations, are available across much of the country to accommodate all the photovoltaic panels that cities will need for a properly decentralised energy economy.

Electricity not needed at any particular moment can be stored in the batteries of electric cars (and then extracted again when necessary via the 'vehicle-to-grid' technology) or put into batteries in domestic homes, such as is being planned on Orkney. Long-term surpluses will be put through electrolysers that won't take up much space and the hydrogen gas can be stored underground, not necessarily in the local area.

If the city's gas grid has converted its pipes to plastic (as many now have), conversion to burning hydrogen in home appliances such as boilers and cookers is entirely feasible. A major research project run by gas industry participants showed that the city of Leeds, for example, could be conveniently converted to using hydrogen rather than natural gas.

LOCAL ENERGY NETWORKS

Getting electricity to the final customer, whether in homes, factories or offices, is a costly activity today. The charges imposed on your electricity retailer by National Grid (the company that handles all long-distance carriage of electricity) and the fourteen local distribution monopolies now represent a quarter of the total domestic bill. These costs have risen substantially in recent years and will rise further in the future.

It would make good sense to give control of energy distribution within a city or a town to the municipal utility, such as the one in Nottingham. One reason that the existing private monopolies should be replaced is that they have increased customer bills by a disproportionate amount, perhaps because they are able to outwit the government's regulator. Even more importantly, they will find it difficult to be effective proponents of the move to local electricity generation, because such a transition threatens their financial future. Our existing electricity distribution companies make money from increasing the amount of electricity transported across their networks. In contrast, we want to ensure that as much possible is consumed close to the point of its production.

The change to local ownership cannot happen overnight. The businesses running the local distribution networks are sophisticated enterprises that are repositories of engineering knowledge. It would be foolish to take them over without very careful planning. Nevertheless, it makes immediate sense to require these cumbersome monopolists to sell parts of the networks for experiments in municipal ownership.

The difficult question is whether the proposed municipal utilities in the UK can successfully both decarbonise energy use and cover their costs. In the UK, solar power can directly compete with other generation methods, such as gas, but Nottingham would also have to pay the price for the conversion to hydrogen of the energy that it needs to store. We'll need to prove it works financially.

By generating or purchasing all of its own electricity, and having the rights to distribute it or store it for later use, a municipal utility should be able to offer competitively priced electricity, and probably heat, to local households. In addition, perhaps local not-for-profit companies should run the local transport networks (and convert them to electricity as fast as possible) and improve the quality of housing, reducing heat needs in winter. Municipal ownership will enable us to push much more quickly towards a low carbon future.

BACK TO ORKNEY

Alongside its investments in renewable hydrogen (see Chapter 1), Orkney is working on the digital infrastructure that a modern energy system needs in order to run on 100 per cent renewable electricity. The operator of the local electricity grid is putting sensors across the network that will enable electricity supply and demand to be balanced second by second. Alongside smart machines that can be turned up or down to match the availability of electricity, the Orkney project will install batteries in 500 homes and 100 offices in order to allow the grid to send surplus power into storage or extract it when the wind suddenly drops. The scheme will encourage the purchase of electric vehicles and help develop the reverse electricity flow of vehicle-to-grid. (More on 'vehicle-to-grid' technologies in Chapter 4.)

A further innovation that my interviewee Adele Lidderdale discussed with me was that the islands intend to pioneer a technology that allows individual homes or businesses that have surplus electricity to sell power to their neighbours. The 700 or so tiny turbines on farms produce 20 per cent of the islands' electricity requirements when the wind is blowing hard. At the moment, the energy coming from a micro-wind turbine on a farm will flow back into the electricity network if the house isn't using it all, and the owner will see little return. The Orkney experiment will allow a neighbour to buy that energy at a price that is lower than that from conventional suppliers.

The proposal to allow power sales to neighbours is based around what is called 'blockchain', a software that can account for tiny transactions. If the sun starts shining and my roof produces electricity for a few minutes that I don't use, my neighbour can buy that power and pay for it immediately. This can all happen automatically using modern digital technologies. Peer-to-peer electricity trading is a hugely important part of the transition to a genuinely local energy system.

What the local Orkney groups are doing is trying to use their own energy resources, rather than buying from the mainland. Among the many beneficial consequences, the cost of living on the islands will fall. More money will be retained in the local community. The economy will be able to diversify away from pastoral agriculture. Skilled jobs in such occupations as wind turbine maintenance and running the local 'smart' grid will multiply. Before starting work at the local council, Adele Lidderdale specialised in sustainable development for rural areas and she tells me the work on Orkney is an extraordinary chance to show that the transition to a 100 per cent clean energy system can benefit groups at the periphery, geographic and financial, of today's economy.

The UK and the rest of the world should watch this experiment closely and benefit from the knowledge gained in the next months and years.

MICROGRIDS AND VIRTUAL ENERGY ISLANDS

Around the world, hundreds of similar ventures show the value of pushing the control of our electricity networks out to local areas. Orkney already demonstrates that, by matching supply and demand in small areas, abundant renewables can become even more cost-effective. Orkney was pushed into experimenting because of the limited capacity of its connection to the mainland grid.

Many islands around the world are following a similar path. Whether it be Hawai'i or the Isles of Scilly off the

Cornish coast, the focus is on developing energy self-sufficiency by using advanced digital technologies, such as smart meters, to reduce dependence on energy from outside the area. Other places are beginning to notice these lessons. Many schemes now focus on trying to make 'virtual islands' in local areas. These also try to align the usage of electricity to its availability within a specific town or even smaller area. Often called microgrids, these schemes operate in areas such as university campuses or remote areas with poor connections to the standard electricity grid. But there is no reason why the Nottingham city utility couldn't also run its system as a virtual island, maximising the consumption of local energy.

The UK utility Centrica, which owns British Gas, is trialling various new technologies in Cornwall that will allow the area to mimic an island and reduce the need to upgrade the overstretched main electricity link bringing power from the rest of England to its westernmost points. Similarly to Orkney, the ability of businesses to connect renewable energy supplies profitably to the grid is being constrained because of an inability to export power.

Centrica is employing a technology developed in the US by a business called LO3. The LO3 microgrid in Brooklyn, New York, is often seen as a model for others around the world. There, a variety of homes and businesses buy and sell electricity from each other. A cinema with solar panels supplies power to a home down the street, while a bakery imports from a battery in the house across the road. Buildings with smart meters record electricity usage every second and communicate the information to the LO3 network. This enables management of supply and demand and settlement of bills for the sale of electricity by one participant to another.

If demand is temporarily too great, an effective microgrid system can turn off or turn down flexible uses of energy, such as electric vehicle charging or air conditioning. Members of the Brooklyn microgrid can buy and sell electricity with each other automatically over a blockchain network. Customers are alerted when the electricity price is cheap, so that they can carry out their energy-intensive activities as inexpensively as possible. Or, when it is scarce, they can release power from their batteries in return for a good price. Advances in sensors mean that every use of electricity around a building – or a town for that matter – can be monitored and controlled to help keep the local network in balance, if necessary by turning appliances up or down.

The rules protecting the monopolists supplying our electricity in the UK mean that peer-to-peer trading does not actually reduce electricity bills. This urgently needs to change, to encourage households and businesses to install solar power or small wind turbines and sell their power to neighbours, as is proposed in Orkney.

An additional advantage is that microgrids can isolate themselves from the wider network and operate independently. This is becoming more valuable because of the increased frequency of extreme weather. In the event of a storm that knocks out the rest of the region, the little grid can continue to provide power, perhaps focusing it on the local medical facilities or other vital services.

Microgrids and virtual energy islands are both really good ideas. They help provide energy independence, encourage the growth of renewables and allow buyers of electricity to financially benefit from the surpluses of others. There's much remaining research to be done to ensure that they work effectively, but a new project in Oxfordshire is trying to demonstrate how the ninety main renewable energy sites in the county can be combined digitally to minimise the need to import electricity and to ensure that each wind, solar and small hydroelectric power site gets the most value from its production. The scheme seems practical and cost-effective.

DETACHING FROM THE GRID

In countries where energy islands are most economically attractive today – rural Australia being a good example – it already almost makes sense for communities to detach completely from the network to avoid distribution charges. As one Australian developer of a technology that competes with LO3 put it: 'In the longer term, we will begin to see more regional Australian towns disconnecting from the main grid and establishing smaller microgrids which are a mix of distributed solar and battery storage ... To think we can continue with the same model of energy supply and get materially different outcomes in terms of cost, safety, reliability and carbon intensity is a folly – we need to be adopting a fundamentally different approach if we're going to get a fundamentally different outcome.'

This is why I am proposing a transfer of control over the local distribution network in the UK needs to be vested in the municipal utilities that will run our energy supplies. We will not get rapid change, otherwise. Of course, these new public utilities will want to import electricity at times. They will therefore still require access to the central backbone provided by the national grid, which will bring electricity from large-scale onshore and offshore wind to our towns and cities.

CHAPTER 3

HOUSES FIT FOR PURPOSE

We need effective insulation of homes and to convert gas boilers to run on hydrogen

Homes produce another 15 per cent of the UK's domestic emissions, mostly through the burning of gas in central heating boilers. The poor quality of the nation's housing means that improving insulation and air tightness is an urgent priority. Heating domestic properties requires almost as much energy as the total UK electricity demand. This is costly for householders and disproportionately affects the less well-off.

The weak and ineffectual programmes instituted by successive UK governments to reduce domestic heating needs have not worked. Gas demand is now tending to rise, not fall. The plausible, indeed blindingly obvious, way forward is a mass, street-by-street programme of 'deep refurbishment' of Britain's elderly housing stock.

For those homes that cannot be significantly improved, a conversion to hydrogen is necessary. To prepare for this, all new central heating boilers need to be switchable between natural gas and hydrogen fuel. And the small number of homes with effective insulation can be switched to electric heat pumps.

THE SCALE OF THE PROBLEM

Electricity use and heating in homes accounts for almost a third of the UK's total energy requirements. Their share of greenhouse gas emissions is somewhat lower, because of the increasing low carbon sources of electric power, but domestic energy use is still directly responsible for about one seventh of our greenhouse gas emissions. This is almost all for heating rooms and water.

The UK has very poorly insulated homes by north European standards, partly because our homes are older than in most other countries and suffer from lamentable air tightness. Although some new homes have better insulation than in the past, many still have high heating needs because of weak regulations and very poor construction standards.

To achieve net zero, we will need to completely wipe out all emissions resulting from our home energy use. It is sometimes tempting to think that domestic heating is such a difficult problem that we should focus on other sources of greenhouse gases first. But, as we will see later, several other sources of carbon emissions represent problems just as intractable. We have little alternative but to focus on improving homes, not least because better housing adds to the standard of living in many different ways, including improving public health.

We can move towads zero carbon housing in three ways:

)) Shift all energy needs in the home, including heating and cooking, to electricity. If we produce 100 per cent renewable power, and this is the only energy a house uses, then we have zero emissions from domestic housing.

)) Continue to use gas, but switch away from methane (the prime ingredient of natural gas) to hydrogen that is made from renewable electricity (see Chapter 1).

)) Insulate all British homes so well that solar panels on the roof cover the total annual demand for energy.

Or we can pursue a mixture of all three options. Some houses could be converted to electric heat pumps; others to hydrogen heating; and some to net zero energy use after deep refits. In new homes we might decide to ban boilers entirely and rely on excellent insulation combined with electric heating (the Netherlands is pursuing this route). We could allow badly insulated older properties to be heated by electric heat pumps, supplemented by hydrogen heaters for the coldest weather.

Whatever the mixture of measures, the speed of change will have to be unprecedented. A million houses a year will need to be upgraded to achieve carbon neutrality by 2050. That is almost 4,000 homes every working day. Which is a challenge, of course, but also an opportunity to create a new industry that will not only address the climate crisis but provide large numbers of jobs and significantly improve the housing conditions of people across the whole country.

SHIFTING HEATING TO ELECTRICITY

At present, a small number of UK homes are heated by electric radiators, often in the form of storage heaters that turn cheaper overnight electricity into heat for later release. A more effective alternative may be to use electricity to run a heat pump that heats water in a radiator system. Heat pumps can be remarkably efficient – one kilowatt hour of electricity may provide three or four kilowatt hours of heat into a house – and, although electric power is currently about four times as expensive as domestic gas, greater efficiency can mean bills are not much higher. Many European countries, notably France, use this technology extensively.

Moving to electric heat pumps looks the easiest way of cutting carbon emissions from home heating, and is recommended by the UK government's advisers at the Committee on Climate Change. In well-insulated homes, heat pumps are the logical choice. However, we shouldn't underestimate how difficult it will be to achieve a successful transition based purely on electric heating.

The first problem is that heat pumps installed in the UK have not achieved anything like the efficiencies claimed. A lack of skilled installers, combined with poor home insulation, means that electricity bills for heat pumps are often far higher than the previous ones for gas. Moreover, many pumps do not deliver sufficient heat to the house's radiators, making it difficult to maintain a good temperature in winter.

The second issue is that the lowest efficiencies for heat pumps will always occur on the coldest days. At these times, it would sometimes be better, and cheaper, to heat a home using a standard electric radiator. Aside from the question of cost to the homeowner, we also need to consider the demand on the electricity network. Gas demand on the very coldest days can be five times as much as the need for electricity.

A further problem is that heat pump installations are expensive, and usually require replacing all the radiators in a house. It will be simpler, and probably less than half the cost, to switch to hydrogen boilers.

HYDROGEN GAS FOR HEATING

In large parts of the UK, the gas network can already carry hydrogen, and by 2031 this will apply to the whole country, as all the iron pipes on the gas network will have been replaced by polyethylene. This will allow safe delivery of hydrogen gas, made from surplus electricity generated at times of excess renewables.

Very little infrastructure will need to be changed, although it will be necessary to invest in hydrogen storage capacity. Existing gas boilers, and other appliances such as cooking stoves, would have to be replaced in homes on the hydrogen network. But the Institution of Engineering and Technology recently said that 'Initial investigations have shown that hydrogen boilers can deliver comparable levels of performance to natural gas for a similar cost. Boilers and appliances can be designed to be "hydrogen-ready", i.e. operating initially on natural gas with conversion to hydrogen at some later date.'

At some point – probably during summer when central heating demand is absent – portions of the existing gas network could be switched to carry hydrogen instead. This switch is a monumental task and would be done over several years, but it is not too dissimilar to the switch from natural gas to 'town gas' in the 1960s. (The old town gas, by the way, was made from coal and was about 50 per cent hydrogen.)

Almost 2 million central heating boilers are sold in the UK each year. The easiest policy for the government would be to mandate that all boilers sold from, say, 2022 will have to be switchable to hydrogen when the gas supply changes. If this was mandatory, the transition to a position where all boilers are hydrogen-capable would take around fifteen years, but a sensible scheme would offer a mixture of incentives and regulations to speed things up.

In the meantime, the UK could follow other European countries and allow a small amount of hydrogen to be mixed into ordinary natural gas. Addition of hydrogen is effectively banned in the UK, but other countries allow up to 20 per cent. This would be an early and large market for green hydrogen produced from surplus electricity. Experiments currently being started by gas network operators eager to push the switch to hydrogen will confirm the feasibility and safety of this switch.

INSULATE HOMES SO WELL THAT THEY USE VIRTUALLY NO ENERGY

The UK government has tried to improve our woeful housing stock over the last couple of decades, initially with programmes focused on insulating cavity walls, improving roof insulation and installing new and more efficient boilers. These had some limited success in reducing customers' bills and cutting emissions. More recently, a convoluted and expensive scheme proposed to lend money to householders to carry out improvements. It was a disaster, costing over £200m, and 'failed to deliver any meaningful benefit', according to a report by a government watchdog.

Ad hoc refurbishments, whether organised by government or done privately, tend not to produce substantial benefits to the overall heat retention capability of houses, even though they can cost tens of thousands of pounds. 'Improve home insulation' is a powerful rallying cry from leading green activists, but achieving significant reductions in heating requirements has proved challenging. Houses made with solid stone or brick walls, of which there are about 8 million in the UK, have been particularly difficult to revamp.

Since 2013, energy saving measures sponsored by government, but usually implemented by the main energy companies, have been installed in about 2 million homes in the UK. That represents about 1 per cent of all homes each year, meaning it would take a century to improve all properties. The results have been unremarkable, at best. Government data suggest better loft insulation typically results in a 4 per cent improvement in gas consumption in a home, with cavity wall insulation only slightly better, at around 7 per cent. These aren't good enough for a country aiming for zero net emissions.

Most disturbingly, total gas consumption across UK households has actually risen by a few per cent in recent years. (This is after adjusting for temperature variations between different years.) Gas use had fallen by about a third from 2005 to 2014 as a result of better appliances and energy saving measures, but it has since increased across various types of housing. The only possible conclusion is that ad hoc individual energy saving measures, as promoted by recent government schemes, have reached the limit of their effectiveness. If we are to radically improve the energy consumption of British housing, another approach is necessary.

'DEEP REFURBISHMENT'

A short row of terraced housing not far from the centre of Nottingham (the city is an interesting example of many different energy initiatives) was the first in the UK to benefit from extensive retrofitting managed by a Dutch not-for-profit organisation. Originally built to very poor insulation standards, these homes used to cost around £1,350 a year just to heat. That's roughly twice the UK average. Some tenants left their houses unheated to save money.

Now, the occupants pay a fixed annual fee for their energy so long as they don't exceed an allowance for electricity usage; the cost is less than half what they used to pay. The houses are warm and comfortable and their external appearance has been vastly improved. The gas supply has been disconnected and replaced by an efficient communal heat pump that pipes hot water to all the homes. Hugely enhanced insulation of the exterior walls and vastly improved windows help keep the buildings at a standard temperature throughout the year. New roofs on each house are fully equipped with solar panels that provide much of the power, and a battery ensures most of the electricity produced is used in the development.

The Nottingham homes were costly to refit. The total bill for the extensive works on each house was about £90,000. Even major savings on the heating bills wouldn't be enough to justify expenditure of this amount of money. But the heating and other problems of these homes were so serious that the housing association that owns the properties was otherwise intending to demolish and then replace them, at a much higher cost. So, in a sense, the refurbishment saved money. A thirty-year guarantee from the construction company adds to their value. And, as Nick Murphy, the head of Nottingham's social housing company, has said: 'Tenants in our pilot absolutely love it. Cold homes have become warm, comfortable and affordable to heat.' Unlike single measures, such as loft insulation, full-scale refurbishment of entire houses seems to make huge differences to heating needs, and to the comfort of the people who live there.

Energiesprong, the Dutch company that carried out the Nottingham refurbishment, is now converting 1,000 homes in the Netherlands each year, as well as working in eight other countries. It has a number of UK developments in progress – including a much larger scheme in Nottingham (with 155 homes) that aims to push the cost down to nearer £40,000 a house – a price point that makes it financially irresistible for social landlords to invest in refurbishment. The company claims that it is already able to convert homes in the Netherlands at a price that needs no external subsidy. It is not just the heating bills of the tenants that are reduced, as many of the landlord's costs of repair and maintenance disappear because the houses are no longer damp. For private owners, there is the additional advantage of getting a much more valuable property as a result of the refurbishment.

Crucially, the core elements of the insulation, including the roof and the panels with windows covering the front and back of the house, are made on a production line in a factory. They are then fitted with minimum disruption to the residents. The Nottingham refurbishments were carried out in just two weeks and it may eventually become possible to reduce this to a few days. Compared with the many months usually taken for a substantial home refurbishment using conventional techniques, this is an extraordinary improvement.

In a successful deep refurbishment, the amount of heating needed falls to a small fraction of the previous level, and electricity demand is low after lighting has been switched to LEDs and other improvements are made. The total energy used is balanced by the electricity generated by the PV panels on the roof. As one academic study concluded: 'The Energiesprong approach makes it possible to take a property in one jump to the required 2050 performance' (i.e. net zero).

This isn't quite the case in the Nottingham development because of the limited roof space on each house, but it should be possible in future UK projects, including those being completed at the moment.

SCALING UP 'DEEP REFURBISHMENT'

Energiesprong says that if the UK completed 25,000 full refurbishments over the next five years, it would bring the cost down to a level that made it one of the cheapest ways of cutting carbon emissions. Social housing, which provides about 4.5 million homes, is the obvious place to start, partly because units tend to be more standardised than privately owned houses. Energiesprong says that its approach would work on at least 11 million UK homes – more than a third of the total.

If construction companies could achieve the target of £40,000 per unit, the total cost to get to net zero on these houses will be around £400bn. Spread over twenty years, this is about 1 per cent of UK national income each year. But warm, comfortable and dry houses would have a remarkable effect on our standard of living and on social equity and public health.

Reducing energy use is the first benefit of major refits, such as the Energiesprong projects. The second major advantage is the boost to local employment and incomes. I suspect that there is no better way of enhancing a local economy and community than by a widespread, street-by-street refurbishment. Such an enterprise will pull down costs and build skills among the local workforce. The resulting houses also look fabulous, particularly compared to their tired appearance before the work.

Amongst other benefits, refurbished houses will improve the health of those living in substandard accommodation. One estimate suggests a value of £1.4bn a year from this alone in savings to the NHS.

In order to understand better the social impact of effective refurbishment, I rang David Adams, the technical director of Melius Homes, the company that carried out the Energiesprong project in Nottingham. He is now busy on a much larger set of deep refurbishments for the city and spoke with passion about the value of the work: 'It is transformative in so many ways. It transforms lives, it transforms neighbourhoods, it transforms carbon footprints,' he said. 'Homes that were producing six tonnes of CO_2 each year are now down to a tenth of that amount. And people can afford to turn their heating on.'

HOW SHOULD WE PROCEED?

Electric heating using heat pumps only seems to make sense in a well-insulated house. Most UK homes do not match this description, and small improvements, such as loft insulation, will not make them suitable. When a house has been properly refitted to bring it to high standards of energy efficiency, a heat pump is the answer. But not before. A better approach is to drive forward with a mixture of deep refits for millions of homes, combined with a switch to hydrogen in the gas network and subsidised conversion of gas boilers. This is a low cost way of ensuring that the UK is able to use zero carbon hydrogen as soon as possible.

We shouldn't underestimate the initial difficulties of getting tenant approval for the refitting of their homes, at least until the success of the Energiesprong approach is better known. Nor should we underestimate the costs of further pilot projects before the building industry has driven down costs. But I am reminded of the early years of the offshore wind industry. Costs were at least a five times multiple of the levels promised for the early 2020s. Consistent large-scale support from government enabled the wind companies to pioneer surprising improvements and push costs down to truly unexpected levels. Housing improvements could be equally as impressive. We will need to create an industry that will eventually be able to deep-refurbish tens of thousands of homes a week at costs far below current levels.

It goes without saying, of course, that regulations should ensure that all new housing – and this is not currently the case – is built to the very highest standards. Most studies show that some properties finished today, especially those converted from offices or commercial buildings, have insulation standards that are worse than new homes built twenty years ago, largely because they are exempt from some building regulations. Every home completed today to low insulation standards represents a financial and carbon burden on the future.

Lastly, it's worth stressing again that these two steps towards net zero homes – effective insulation and replacing natural gas with hydrogen – can be controlled and managed by local public enterprises. The local gas networks can be operated by city utilities, and building refurbishment can also be put under the control of social housing departments. The potential benefits to local employment and to living standards are huge.

CHAPTER 4

ELECTRIC TRANSPORT

A fast track to electrification, car-sharing and free public transport

Transport generates more than a quarter of UK emissions. Car use is the most important source, accounting for about 16 per cent; heavy lorries and vans add another 9 per cent and buses and train travel a further 1 per cent. Emissions from international transport (planes and shipping) are significant additions (flights add a further 7 per cent to the UK's emissions), but they are usually omitted from calculations and their particular challenges are covered in the following chapter.

For land-based vehicles, electrification is key and will reduce emissions substantially. Even if the electricity to charge a battery was generated from coal, electric motors are so much more efficient than internal combustion engines that electric vehicle (EV) use cuts greenhouse gases. However, building a new electric vehicle involves major use of energy and natural resources. The UK needs to replace its 30 million existing cars with a much smaller number of vehicles, and to make it simple to rent these for short periods. Just as important, we must improve public transport and facilities for walking and biking. We should try to get cars out of towns and cities.

Most road freight can be switched to electricity. The largest vehicles may transfer to hydrogen fuel cells, which use the gas to make electricity.

THE IMPORTANCE OF ELECTRIFICATION

In the published figures on UK domestic greenhouse gas emissions, transport looms ever larger. It was 21 per cent of the country's CO_2 emissions in 1990 and it is now fully one third. Recent reports have thrown up alarming statistics on emissions from cars, which have gone up, despite the increase in battery and plug-in hybrid cars. The reason is the growth in SUVs, which are hugely energy inefficient and have bucked an earlier trend for smaller cars. Globally, SUVs are now the second most important driver (after new power stations) of the increase in emissions.

The typical British car is driven about 8,000 miles a year. This means that the average new car produces about a tonne and a half of CO_2 a year; older cars generally emit more CO_2, as well as much higher levels of other pollutants. The regulatory drive across Europe to reduce new car emissions has reduced average emissions from new vehicles by about 30 per cent since the turn of the millennium. But the improvement has recently stalled because of the swing away from diesel cars, which produce less CO_2 per mile than their petrol equivalents, and the rise of SUVs.

Cars powered by batteries and without an internal combustion engine are still only a small fraction of total vehicle sales. In October 2019, they constituted 2.2 per cent of new cars bought from British dealers, although this figure represented an almost threefold increase in a year. In Norway, where the government actively encourages EV purchases and provides incentives for their use, electric cars now represent more than half of all sales.

Electric cars in the UK are currently more expensive than their petrol equivalents and the deep savings in running costs appear to be insufficient to push buyers to switch. Are customers being logical? And do we need subsidies and perks like use of car lanes and parking?

❱❱ At today's (late 2019) prices of petrol, the average new vehicle will cost about £1,000 a year in fuel. An electric equivalent, charged entirely at home or work, will come in at around £300. (It would be more if the driver had to pay extra for using public chargers.) Over the fifteen-year lifetime of a car, an EV therefore saves at least £10,000. Electric cars are also much cheaper to maintain and no more expensive to insure. So the current price premium may actually be well worth paying. For drivers who drive much more than the average, this is doubly true.

)) Many people remain worried about the availability of charging points for their long-distance travel. This is understandable, because large areas of the country have limited numbers of the fastest chargers. But this is changing rapidly as commercial businesses such as BP, Shell and the electricity companies install rapid charging points as fast as they can. It makes good financial sense for private companies to do this, so we should be confident that the growth in installations will continue.

)) Electric cars are continuing to fall in price. We don't know whether price parity on the forecourt is one year away or three, but the trend is clear. Volkswagen, which has put more into electrifying its range than any other Western manufacturer, sees battery prices continuing to fall sharply. This cost is the single reason why electric cars won't be more expensive in a few years' time. As a Volkswagen executive said: 'We strongly believe that the tipping point is near, and that tipping point will be price (equality).'

)) Most EV drivers strongly prefer their electric cars to the petrol vehicles they have used in the past. They help sell the benefits of electric cars to other drivers.

Over the next fifteen years, using a mixture of taxes and regulations, we should be able to push almost all new car sales towards EVs. An electric car will save carbon emissions compared to a petrol equivalent. Iain Staffell of Imperial College, London, states that 'on average Britain's EVs emit just one quarter the CO_2 of conventional petrol and diesel vehicles'. This is partly because so much of British electricity now comes from wind, solar, biomass and nuclear. But is simply switching to electric cars good enough?

SHARING AND GIVING UP CARS

One 2015 study suggested that an EV with a good range would have a much higher carbon footprint in its manufacturing process than a petrol equivalent, proposing a figure of about 16 tonnes per car. That's approximately 1 tonne for each year of the vehicle's life. Much of this footprint comes from electricity, which is rapidly decarbonising. However, manufacturing a machine such as an EV, which contains well over 1 tonne of raw materials, will always have a significant carbon impact.

A net zero society may be obliged to move away from the personal ownership of resource-intensive goods, of which cars are by far the most important example. Urban car-sharing clubs are a good first step on this route. but we will need to go further and substantially reduce the number of vehicles on the roads. We should also note that, although electric cars are far less polluting than their equivalents with engines, they still produce large amounts of micro-particulates that seriously affect health. (These arise from the wear on the brakes and tyres.) The fewer cars of all types on the roads, the better for urban health.

Of course, we will only be able to reduce car use if public transport, pedestrianisation and cycling routes are massively improved. As many European cities are finding, if we can improve the attractiveness of alternatives, we can gradually remove the private car from the centre of urban areas. And everybody will benefit from this.

Utrecht, a Dutch city of about 350,000 people, is probably the best known example. There, 60 per cent of residents use bikes to get from their homes to the city centre.

Over a quarter of all trips are made on foot, compared to 19 per cent using vehicles. There are 3,300 shared cars. The city is completing the final construction of a bicycle park with 33,000 places near to the rail station, one for every eleven people living in the city. Every day, more than 100,000 people ride through the streets. Forty-three per cent of all journeys shorter than 7.5 kilometres are by bike, compared to a figure of 2 per cent for the UK.

One study across the Netherlands suggested an increase in Dutch life expectancy of over half a year as a result of the average seventy-four minutes of cycling the population engage in each week. An estimate of the benefits to Utrecht specifically was for a reduction in annual health care costs of $300m, compared to the yearly budget of $55m for cycling infrastructure. So, improving cycling and walking facilities around towns and cities has to be a vital part of any municipal Green New Deal alongside local charging infrastructure for private cars and vans.

FREE PUBLIC TRANSPORT?

In most of the UK, public transport needs major improvements. We should be looking at the growing number of experiments around Europe involving free bus transport. This increases usage and helps revitalise city centres.

The French coastal city of Dunkirk, which made all its bus travel free in September 2018, is a good example. The cost to the city was relatively small, since fares only covered 10 per cent of the system and this money was found by marginally raising a local employment tax. Use of public transport rose sharply and the city is on course for a doubling of the number of passengers by the end of 2020. Bus priority lanes mean they travel slightly faster than cars. But the main impact, according to the mayor, has been to increase economic activity. Shops are busy, people get out more. The mayor said, 'All the towns which have moved to free buses say the same thing – the revitalisation of the city centre, it's the most important effect.'

How might free public transport be financed? In Tallinn, Estonia, it is funded by a share of national income tax, but it could be logical to finance it through a local charge for private cars. This would help reduce congestion and improve transit times for everybody. Another advantage is that raising the cost of using cars and cutting the price of public transport improves economic equality as well as making travel quicker.

An even more radical (and invariably popular) solution is wholesale pedestrianisation. This probably works best for medium-sized towns – ones which you can walk across in half an hour. Examples like Pontevedra in north-west Spain again show how city centres can be revitalised through such initatives. In Pontevedra, the city population has risen (in contrast to the shrinking of nearby towns), traffic accidents have fallen and CO_2 emissions are down by 70 per cent.

CAN WE ELECTRIFY CITY TRANSPORT?
Buses can be highly polluting in urban centres, contributing to increased incidence of asthma and other diseases.

Fortunately, it makes at least as much sense to electrify buses as it does cars. Around the world, we'll probably see a much more rapid shift to fleets of battery-powered public buses than other types of vehicle. Urban buses travel relatively small numbers of kilometres per day, and have to return to the bus station at night, where they can be easily charged.

In the huge metropolis of Shenzhen, China, all 16,000 buses are electric, as are the taxis. The price was enormous, with each bus reported to cost over £200,000. However running costs are dramatically lower, and urban pollution is reduced. Most calculations show that lifetime usage costs for electric buses are lower than their diesel equivalent already, and this advantage will only increase as battery prices continue to fall.

Taxis can also be easily switched, too, particularly in dense urban areas. Although very expensive to buy, the new electric London taxi saves its drivers an average of £100 in diesel costs each week, or over £5,000 a year. The drivers I have spoken to are all happy with their switch from fossil fuels.

With trains, the most cost-effective way of moving to low carbon emissions may well be to convert them to hydrogen, using a fuel cell which converts the gas back into electricity. There are already hydrogen trains working on short journeys in the north of Germany, and plans exist to start using them in other parts of the country. In France, the rail operator has promised to get rid of all diesel trains within fifteen years, replacing them with a mixture of hydrogen and hybrid equivalents.

Some UK train operators will also experiment with hydrogen over the next couple of years, expecting to take paying passengers in 2021. These quiet, low pollution trains have a similar range to diesel and a very healthy maximum speed. The hydrogen, of course, can come from renewable electricity and causes no local pollution when used, unlike the highly polluting diesel trains that dominate Britain's rolling stock.

VEHICLE-TO-GRID BATTERY STORAGE

This chapter proposes that most of our transport needs will be provided by electric transport, whether it be short-distance ferries, buses, electric bikes or cars. All of these will carry batteries, often with huge amounts of electricity storage. Most vehicles are used fairly predictably, whether buses that run from five in the morning to eleven at night, or cars that just drive to and from work or for weekend trips. In either case, most of the time (up to 95 per cent, in the case of personal vehicles) vehicles stand idle. It is then that their batteries can be used as a supplementary source of storage for the electricity grid. At times of surplus, power can be directed to filling up batteries or taking electricity out when supplies are scarce.

These 'vehicle-to-grid' technologies are still in early development but will provide vital support for grid stability. One of the many advantages of batteries as a mechanism for balancing supply and demand is the speed at which they react. Called upon to act by a digital signal from the electricity network, a battery will respond in less than a second, a far quicker reaction than any form of power station.

Reversible batteries in electric cars, vans and taxis could easily provide the spare capacity to enable a grid to weather problems, such as a large wind farm suddenly disconnecting. In August 2019, for example, the UK's national grid lost about 1.4 GW of capacity over a few seconds and the local distribution companies had to reduce power usage by about 5 per cent across the UK. Trains were stopped and hospitals lost power. Just 200,000 vehicle batteries could have avoided the entire problem and the UK will probably have this number within a couple of years.

Across many parts of the energy transition, the UK has already lost out to more advanced competitors. In the case of battery manufacture, for example, the advances made by Tesla and its partner Panasonic have made it impossible for UK businesses, even with innovative university partners, to compete effectively. But 'vehicle-to-grid' software and hardware is still an open field, which Britain could hope to dominate. We need to push governments and research organisations to concentrate their efforts on such technological developments.

CHAPTER 5

FLIGHTS AND SHIPPING

We need to fly less - the hardest challenge for zero carbon. And shipping should run on hydrogen

Greta Thunberg's August 2019 trip across the Atlantic in a racing yacht dramatically highlighted the climate impact of both air and sea travel. Long-distance transport is very difficult to decarbonise and aviation is perhaps the most difficult challenge on the road to a zero carbon society, particularly for the UK. The British population is one of the heaviest users of air travel today, and around 7 per cent of our emissions are from aviation, compared to a worldwide average of less than a third of this figure. On an individual level, just one return flight from London to New York may result in more CO_2 than running a small modern car for a year.

Some good news is that most shipping (which represents 2.5 per cent of global CO_2 emissions – and probably a higher figure for the UK) can probably be transferred to hydrogen. Air travel is much more problematic, partly because the release of pollution high in the atmosphere is more damaging than at ground level, but also because large planes cannot realistically be electrified.

IS FLYING AN INTRACTABLE PROBLEM?

Aviation emissions are rising – and showing no signs of changing course. Globally, at any time of day or night, 10,000 planes are in the air, carrying more than a million passengers. And, shockingly, the UK contributes the highest number of international air travellers: 126.2 million passengers each year, or 8.6 per cent of the world's total (the US is second, with 111.5 million, followed by China with 97 million).

Most scenarios for our carbon future suggest that there will still be many tens of millions of tonnes of CO_2 emissions a year from aviation in 2050. Indeed, the government's Committe on Climate Change states that 'Plausible options for how aviation could become zero carbon, even by mid-century, are lacking' and that they 'expect the sector to emit more than any other in 2050'.

The assumption of most such reports is that the remaining emissions in 2050 will have to be counterbalanced by net CO_2 capture elsewhere in the economy. Some analysts point to the possibility of electric planes, 100 per cent conversion to biofuels for aviation and increases in energy efficiency. However, none of these are likely to make a significant difference to total aviation emissions, particularly given the continuing upward trajectory of flight numbers.

The problem with electric planes is the weight of batteries needed for each unit of power. Even super-efficient batteries have only one sixth of the energy density of aviation fuel. This is critically important, since about 35 per cent of the take-off weight of an plane crossing the Atlantic today is aviation kerosene. Batteries to power such a trip would simply be too heavy. Short-range aviation, such as London to Amsterdam, might be conceivable by 2050, but even converting 400-kilometre flights to electricity won't be enough to dent the usage of fuel by more than a few per cent.

Fuel for planes can be made from biologically derived materials, such as vegetable oils. In theory this could be low carbon, if the carbon in the oil has previously been captured from the atmosphere by the plant. At the moment, these substitutes are considerably more expensive than fossil fuels – replacing aviation fuel with the cheapest biofuel would add $200 per passenger for a London to Sydney flight. But a more critical issue is that biofuels use up agricultural land that could otherwise be growing food or carbon-capturing trees. At cruising altitude a large jet using 100 per cent biofuel (and no flights do so, as yet) would consume the equivalent of 40,000 square metres of palm oil production an hour. Increasing global forest cover (a vital priority – see Chapter 8) while switching from fossil oil to biofuel seems a near-impossible combination.

Substantial increases in energy efficiency are possible. One think tank suggested that aviation could become 30–45 per cent more efficient by 2050 by using new engine designs and improved aerodynamics. This is a useful gain, but continuing growth of traffic would almost certainly outweigh the effect of these improvements.

The only available solution may therefore be to develop sources of synthetic fuel, made from hydrogen and from captured CO_2 (as described in Chapter 1). This will initially involve a cost premium over conventional aviation kerosene, but it should be far less expensive and less burdensome than a massive uptake of biofuel. One estimate is that it will have an initial fuel premium of around 80 per cent, but this should fall if synthetic fuels becomes a significant industry, perhaps dropping as low as 50 per cent. This would mean an added cost of around $12 for a flight from London to Barcelona. This is not a very painful or discrimatory rise and, once synthetic fuels are a reality, the government could make them mandatory for all planes using UK airports.

There is, however, an additional complication. Burning fuel at high altitude produces water vapour and other pollutants. This has a temporary further effect on the climate, increasing the amount of heat that is trapped. Most estimates see this effect as approximately doubling the climate impact of aviation, at least in the short term. So, even if we create a carbon neutral fuel, made synthetically from renewable energy, aviation will still have a major effect on the climate. This needs to be addressed in any solution to aviation emissions, possibly by removing CO_2 from the atmosphere.

OFFSETTING EMISSIONS

What about planting trees or taking other measures to offset the damage caused by flying? Many organisations will take cash for projects around the world that promise to capture CO_2 or reduce the burning of fossil fuels. The most popular involve reforestation of denuded tropical landscapes or planting new woodland in Britain. Others allow you to sponsor the provision of energy-efficient cooking stoves, which help households in developing countries reduce their use of wood or charcoal.

The most innovative offsetting scheme is discussed in Chapter 11 – the Climeworks offer to take CO_2 out of the air and put it into permanent storage by reacting it with basalt rock in Iceland. This persuasive scheme is, alas, very expensive indeed, at present costing $180 to offset a flight from London to Barcelona and around $1,000 for London to New York. If the Climeworks technology rolls out this figure will fall dramatically but, nonetheless, it will remain a high price. By contrast, one of the conventional 'offsetting' companies is happy to charge you as little as £7.50 ($9) for London to New York.

Do such offsetting schemes work? Specifically, if you give some money to a company, will global emissions go down by the quantity specified on their website? With the exception of the Climeworks venture, which would genuinely sequester a tonne of CO_2 permanently, the answer has to be rather sceptical. There's very little independent quantitative evidence of the effect of offsetting, and some projects that are backed by offsets would probably have been developed anyway. However, handing money over for replanting forests, or avoiding deforestation, or promoting low carbon cooking fuels, or solar lamps, does no harm, and might even do some good.

Nevertheless, pending the development and adoption of synthetic fuels, it is almost certainly better to avoid flying whenever possible, and particularly so when one considers the extra global heating effect caused by the emissions in the high atmosphere.

SHOULD WE AVOID FLYING?

It may seem pointless for individuals to decide to use less polluting forms of transport than aviation. One person's flights make little difference to overall carbon emissions. The planes will fly anyway, perhaps with one fewer seat occupied. But this is unimaginative of us. Flying is generally the most climate-polluting activity we engage in. And, if we make a decision not to fly, it affects the attitudes of those around us.

By avoiding flying, we are signalling that we view the climate crisis as a serious threat to the future of humanity. This has a beneficial effect on politicians, who feel emboldened to act, and also obliges the airlines to begin to develop low carbon fuels at an increased pace in order to save their businesses. The 'flight shame' movement has taken off fastest in Sweden, where the Nordic airline SAS recently announced a 2 per cent fall in traffic in 2019 and blamed climate activists for the impact on its revenues. We can expect SAS to be a leader in pushing for low carbon fuels.

Those who oppose flying are often portrayed as anti-business. Airports, in particular, are eager to assure us that our prosperity is dependent on increasing the number of flights so that businesspeople can travel quickly, cheaply and easily. And their lobbyists are successful. Witness that both the Conservative and Labour parties supported the expansion of Heathrow Airport at the same time as voting to declare a climate emergency.

The truth is that a remarkably small fraction of all air travel is for business purposes. At Gatwick Airport, for example, only 14 per cent of travellers are on business. Heathrow has a higher percentage (26 per cent), but that number is down from around 34 per cent a decade ago. If anything, businesspeople are flying less. Modern communications technologies, such as video-conferencing, make business travel far less necessary.

The growth in aviation is almost all coming from tourism and family visits. And it is a small minority who take most of the flights. A recent survey established that 70 per cent of flights from the UK are taken by just 15 per cent of the population. Indeed, just 1 per cent of the UK population account for 20 per cent of flights.

As to the expansion of Heathrow, this is clearly at odds with our ambitions for zero carbon emissions. One step that would be easy for the government to take would be to place a carbon tax on domestic flights (more on this in Chapter 10) and to use its proceeds to subsidise train and bus travel, in order to remove any financial incentive for short-haul flights, when other forms of travel are easily available and take no more time. The government, too, should examine carefully the benefits of transit passengers using the airport as a hub.

So what should be the right route for the UK? Ideally, we should all fly less – or not at all. But for some people this is well-nigh impossible. We therefore need a mixture of additional measures. First, the UK should be at the forefront of efforts to develop synthetic aviation fuels from hydrogen and captured CO_2 (perhaps putting pressure on the airlines to make this mandatory by, say, 2030, or to face serious taxation). Second, we need to develop realistic and auditable carbon offsetting measures and to develop genuine offsetting on the Climeworks model (again, the ariline industry could be mandated to do this). Third, we should invest in better long-distance transport to reduce the need for aviation inside the UK (or around Western Europe).

SHIPPING – ANOTHER ROLE FOR HYDROGEN

International shipping is only beginning to recognise its responsibilities for CO_2 and the horrifying amount of other pollutants that large carriers emit. The ships of the world's biggest cruise line emit ten times more sulphur oxides in their voyages around Europe than all the continent's cars. This appears to have had serious impacts on health in some cities, such as Barcelona and Venice. The reason for the sulphur pollution is that large ships have been allowed to burn the lowest quality oils and had no responsibility for cleaning up exhaust gases.

This is gradually changing. However, molecules of sulphur oxides in the atmosphere act to reduce global warming, so improving pollution controls will raise the shipping industry's net contribution to global heating. The solution is probably switching short-distance shipping to batteries and swapping longer journeys to hydrogen combined with fuel cells.

Why is this the right route?

For short distances – for example, a ferry between the Isle of Wight and Portsmouth – a battery and an electric motor can easily provide sufficient power. Many similar short routes in the Nordic countries are preparing to shift to electricity as the source of power. The first battery ferry, appropriately called *Ampere*, came to Norway in 2015 and operates a twenty-minute route across a fjord between two villages. It recharges for just ten minutes after each crossing, which can transport as many as 120 vehicles and 350 passengers. Siemens, which made the electric equipment for the ferry, says that the electricity for each crossing costs less than £5.

There are probably only about 200 battery-powered vessels in the world today. However, Norway sees an industrial opportunity making electric vessels and has encouraged local ship operators to invest in battery-powered ships. They are cheaper to operate, more reliable and very much quieter. We can push for a faster conversion to electric short-distance shipping, but hydrogen may be better for other types of route, as Adele from the Orkney council outlined in Chapter 1.

The idea is that the existing huge engines in ships would be taken out and replaced with liquid hydrogen storage and fuel cells that turn the gas into water and electricity. Electric motors would then drive the ship, much as they do in a battery-powered vessel. This is costly, but not prohibitively so.

We are at a very early stage in the transition away from marine diesel, and the first fuel cell ships have yet to be put into commercial service. Concerns over safety and the cost of installing bunkering of hydrogen around the world are seen as obstacles to a rapid switch. Nevertheless, hydrogen looks a fully feasible alternative to today's fuels. France and Norway are taking the lead, as they have done with battery-powered boats. In 2021 a commercial hydrogen ship will enter service west of Oslo on a trip taking about thirty minutes. In San Francisco Bay, an eighty-person ferry will commence operations in 2020, taking commuters to and from work.

These are tentative first steps. The world has more than 50,000 ships that ply between countries, ranging from huge container ships to smoky small freighters that move around the Baltic. Shifting these boats from marine diesel to hydrogen is a huge task, but there are few – if any – alternatives if we want to move to low carbon transport. And changes may happen quicker than we imagine, perhaps in Asia. One senior executive at a Japanese business specialising in power transmission for ships recently asserted that 'hydrogen's uptake in maritime will take a lot of people by surprise, with developments moving ahead quickly – in both Europe and Asia, particularly China. There are obvious challenges, in terms of production, bunkering and other infrastructure, but demand will work as a powerful driver to help industry overcome these issues.'

CHAPTER 6

SUSTAINABLE FASHION

Without big changes, clothing alone
will stop us achieving net zero

Fashion represents about 3–4 per cent of the UK's carbon footprint, though it is not always accounted for in national estimates. Most of the emissions of greenhouse gases take place in the countries where raw materials are made or the clothing manufactured. Both cotton and polyester – the dominant materials – have substantial environmental problems. Polyester has a high carbon footprint, while the cultivation of cotton and its processing into fabrics causes major pollution and uses vast quantities of water. The fibres of both cotton and polyester are shortened each time they are reprocessed, making full recycling very difficult. Virtually no clothing is fully recyclable and very little is reused for a second time.

The British buy more clothes than any other country in Europe, encouraged by low prices in UK shops, and the number of times a piece of clothing is worn has been falling. After purchase, we typically wear an item less than 100 times, and then dispose of it just over two years after purchase, even though it is usually still in good condition. So with no immediate solution to the problem of fashion's high carbon footprint, the best policy is simple: we need to buy fewer garments and keep them for longer.

THE LINEAR ECONOMY AT ITS WORST

Nothing typifies the world's environmental problems more than the clothes we buy. Whether using cotton or a synthetic fibre such as polyester, clothes manufacture requires the emission of large amounts of CO_2 and results in substantial flows of waste. Although some clothes are re-sold in second-hand shops – and a diminishing amount is exported to Africa – most clothing is thrown away while still perfectly wearable. Virtually none is recycled; less than 1 per cent of clothing is made into some form of new clothing after being thrown away. Only about 13 per cent is reused in any way, mostly for making into bedding or cleaning rags. Most fabric ends up in incinerators or landfill.

This is the 'linear' economy at its most destructive: it starts with severe environmental pollution, largely in less prosperous countries where the clothes are made, and is followed by a limited period of use and then by a serious waste disposal problem. And it is getting worse.

Worldwide, the average number of times an item is worn before being thrown away is a third less than fifteen years ago. The clothing industry pollutes in many additonal ways, too, through the degradation of water supplies and the pollution caused by agricultural fertilisers.

Clothing thus represents a significant environmental concern. It is the largest part of our carbon footprint after running the house, using a car, taking flights and eating food. The 18 kilos or so of clothes that the average Briton buys each year cause almost half a tonne of greenhouse gas pollution. (We purchase large volumes of clothes partly because our retailers have been so successful at bringing down the cost in the shops.)

Some of this carbon footprint comes from washing and drying clothes during use, but two thirds arises from the processes involved in making them. Across the world, the fashion industry is responsible for about 1.2 billion tonnes of greenhouse gases a year, equivalent to the combined figures for aviation and maritime shipping. One forecast suggests that the total volume of clothing produced is likely to reach 150 million tonnes by 2050, almost three times the current level. Without major changes, clothing alone will stop the world moving to net zero.

AN INDUSTRY THAT WANTS TO CHANGE
All of the world's largest fashion retailers are fully aware of the impact of their industry on the world's environment. Indeed, compared to the oil and gas

companies, they often show a refreshing honesty about the problems their industry causes. Each year sees yet more detailed reports from research institutes, usually paid for by the large global retailers, which intelligently examine the dilemmas facing the industry. But the recommendations of this research usually focus on small changes to production processes to help avoid local pollution rather than offering suggestions on how to reduce climate emissions to close to zero. They are not yet able to suggest a way of developing anything approaching a 'circular' textile supply chain. This would involve making clothes and then recycling the material into a new product after the original purchaser has disposed of them.

This incompleteness is understandable. We don't have sight of a route to a carbon-neutral clothing industry. Moreover, the fashion business is largely focused on persuading customers to buy a stream of new, inexpensive products, rather than paying more for a limited number of clothes that might last for decades. A report from international consultants McKinsey reminded us that 'Zara offers 24 new clothing collections each year; H&M offers 12 to 16 and refreshes them weekly. Among all European apparel companies, the average number of clothing collections has more than doubled, from two a year in 2000 to about five a year in 2011.'

Most 'sustainability' initiatives from the global textile industry, which now arrive almost monthly, are ineffective, confusing to consumers and simply disguise the underlying problems. Progress is certainly being made on reducing the pollution from textile factories, and possibly in ensuring that pay and conditions for clothing workers are improved. But the deep-rooted problems of high carbon emissions, degradation of cotton fields and truly staggering use of water remain largely unchanged.

COTTON, SYNTHETICS OR WOOL?

It is often assumed that the environmental problems arising from textiles can be avoided by switching from oil-based materials, such as polyester, to natural fibres, particularly cotton. In reality, such a change would achieve little. The processing of cotton fibres creates major chemical pollution. Furthermore, the fabric cannot be fully reused, because the cotton fibres are shortened in the recycling process. Old cotton fabrics will probably always need to be supplemented by a large amount of virgin fibre when making new clothes.

Polyester is slightly easier to recycle, but deteriorates each time it is reused. It also needs to be supplemented by virgin material in order to be suitable for making into a new fabric. This is one of the reasons why virtually no clothes are truly recycled today.

Over 50 per cent of UK clothes by weight are made from cotton, a figure much higher than the worldwide average of around one third. (Worldwide, polyester provides over half of all clothing fabrics.) Cotton is often touted as environmentally more benign than the use of oil-based textiles such as polyester and polyamide. That conclusion may well be wrong.

>> Cotton production uses about 6 per cent of the world's agricultural land. By 2030, according to one estimate, another 1.2 million square kilometres (more than a third more land) will be needed to provide the space for cotton and other natural textile fibres. The conflict with food production and the need to increase forest cover is obvious.

>> Cotton uses vast amounts of water, often in areas of the world with inadequate supplies. A kilo of cotton clothing – perhaps a T-shirt and a pair of jeans – might have required over 10,000 litres of fresh water to grow. That's more than the average person drinks in a decade. As an increasing fraction of the world moves into recurrent water shortages, wasteful cotton cultivation is difficult to justify.

>> Cotton farming also requires substantial use of pesticides and, in most places, high levels of artificial fertilisers. One source estimates that it is responsible for 24 per cent of insecticide use and 11 per cent of other pesticides.

>> Growing organic cotton reduces its environmental impact, but doesn't help with land use requirements. In fact, since the productivity of organic cotton is lower than conventional cultivation, it can be argued that it crowds out even more food growing or reforestation.

>> The production of textiles from raw cotton involves the discharge of hazardous chemicals into water courses.

So cotton is generally bad news. But unfortunately polyester seem to have an even higher carbon footprint. One study suggested that a polyester shirt results in over 5 kilos of emissions compared to just over 2 kilos from a cotton equivalent.

Consumers face a difficult decision: should we buy a partially recyclable polyester garment with a high carbon footprint or a cotton equivalent that has added to the world's other major environmental challenges?

And, for those of us in colder countries, is wool an environmentally acceptable alternative? Unfortunately, the answer is no. Although wool's manufacturing processes are not especially polluting, the footprint of woollen clothes is dominated by the methane produced by sheep. Sheep's wool (and that of cashmere goats) is responsible for more global warming gases than other fabrics. A polyester-based fleece made from recycled plastic drinks bottles (the US company Patagonia is a pioneer in this process) is a far more ecologically friendly alternative to a new woollen sweater.

What about other fabrics made from natural materials? Is hemp or linen clothing, for example, any better than cotton? These sources do produce long-lasting fabrics with a smaller ecological footprint (but they are also very much more expensive than the alternatives). Bamboo, which requires few agricultural inputs to grow rapidly, needs substantial chemical processing in order to be turned into a fabric.

The unavoidable conclusion is that, to reduce our own impact on climate change and wider environmental problems, we need to reduce our purchases of new clothes. Ever-increasing cotton production is incompatible with our need to return large areas of the globe to forest while maintaining adequate space for agriculture. And, until we have found ways of making ethane (the key ingredient in plastics used for clothing) from fully renewable sources, we should not be transferring to polyester textiles.

WHAT ELSE CAN WE DO?

As consumers, we have many means to reduce the global impact of our clothes choices. The simplest is to buy fewer, higher-quality, clothes, designed for longevity. Many items we buy today are not designed or manufactured to withstand repeated wearings. The 'fast fashion' trend has been built on selling clothes to be worn on a few occasions before being thrown away. We cannot entirely blame the retailers for this wasteful and destructive feature of the modern high street. Fast fashion has developed because British shoppers are eager to buy impossibly cheap T-shirts and other disposable items.

Moreover, one study showed that less than a fifth of clothes are disposed of because they are worn, stained, damaged or have lost shape. The other 80 per cent are discarded because they didn't fit any more or because the owner didn't like them. Clothes in good condition need to be reused, and sold on. We could all do well to do more clothes shopping at vintage and charity shops, as well as on sites like eBay and Depop, the latter an encouraging new teenage trend.

To get a sense of the scale of the UK's recycling efforts, I asked Fee Gilfeather at Oxfam how much clothing is diverted from landfill. She told me that her charity's shops process about 14,000 tonnes a year. Some of the clothes are sold in Oxfam shops, some are exported or used for other purposes such as making bedding, and the least useful items are incinerated in a power plant to make electricity. In total, these alternative routes reduce landfill waste by about 5 per cent. Fee estimates that total sales of clothing in charity shops are rising at a fairly rapid 7 per cent a year (though, of course, so are sales of the original items). So the reuse of clothing is becoming more important, but it is still a small fraction of total sales.

We might also look more at short-term rental and sharing. Across a variety of different products, notably cars, we no longer always want to own the things we use and are happy to pay to rent them for short periods. A quick calculation shows that if we typically wear an item of clothing 100 times, and keep it for three years, we are using it for about 5 per cent of the time we own it. Specialist clothing, such as dress for formal occasions, is worn far less and should generally be rented. This happens already to an extent, but could be much more widespread with items such as ball gowns and wedding dresses.

We also need to relearn the habit of repair and reuse. Round the corner from where I live, Oxford Alterations repairs and resizes clothes of all types. This small business operates in a building alongside other carbon-reducing activities such as an enterprise that rents out consumer appliances, like pressure washers, for short-term use. For clothing that no longer fits, or has been damaged in use, Oxford Alterations can make inexpensive changes so you can continue wearing the garment. They even hold workshops to help you make your own clothes.

Businesses like this will be central to the development of the 'circular economy', as well as producing good-quality skilled employment for local people. 'Fixing something we might otherwise throw away is almost inconceivable to many in the heyday of fast fashion and rapidly advancing technology,' says Rose Marcario, the CEO of Patagonia, one of the most sustainable fashion companies in the world. 'But the impact is enormous. As individual consumers, the single best thing we can do for the planet is keep our stuff in use longer.'

A more profound change might come from making different assumptions about clothing. Rather than using bright new garments as our way of showing off, or feeling good about ourselves, we should look to buy clothes that might instead show off our conscience and reliability. This will be a tough change for most of us to make.

In order to make clothes last longer, we need to wash them as little as possible – don't tumble dry and don't iron are the key new rules. And these rules apply also to emissions, as about a third of emissions related to clothes come from looking after them. Tumble drying is particularly energy-consuming (though this should become less important, of course, as we move away from using fossil fuels for electricity generation).

Another good reason for not frequently washing clothes is to avoid microfibre shedding from synthetic fabrics, which often end up in the sea. Each time a garment is washed, vast quantities of tiny fibres are released – 100,000 per wash for a new piece of polyester – and these can be ingested by tiny marine creatures. Washing polyester clothes in a Guppyfriend bag, promoted by manufacturer Patagonia, can dramatically reduce the number of fibres lost. In the longer term, washing machine manufacturers need to improve their filtering systems.

FUTURE CLOTHING

A small number of suppliers already use fabrics with a low carbon footprint – the most important being cellulose, which is derived from wood. Making a cellulose textile requires large amounts of energy but it is far less polluting than cotton. The amount of land needed for the wood is a small fraction of the area required to make cotton, and woodlands don't need extensive chemical fertilisation. Cellulose can be recycled, too, although (as with cotton and polyester) the fibres shorten each time it is reused.

One generic name for cellulose material is 'Lyocell'. Tencel is the best known brand, made by the Austrian firm Lenzing. Some major retailers in the UK stock clothes made from this product, or as part of a mixture with other materials, such as linen. These range from shirts to denim trousers. Tencel garments tend to be more expensive than cotton, but the premium is not huge.

Lenzing is worth singling out here as an example of best practice in the fashion industry. Aware of the high energy use in making Tencel, it has committed to only use low carbon sources in its factories and aims to halve emissions from its production processes by 2030. It also recycles all the chemicals used in its factories, claiming a recovery rate of 99 per cent for its solvents. It does, however, use strong dyes, because the fabric is less good at accepting colour than cotton.

Tencel has many attractive characteristics aside from its environmental performance. It is soft, breathable and resistant to wrinkling. It doesn't encourage bacteria growth when moist, and so is suitable for sportswear.

THE PROBLEMS WITH TAKING ACTION

The key recommendation in this chapter is that the best way to reduce emissions from clothing is by buying less and wearing clothes for longer, rather than relying on the fashion industry to suddenly develop genuinely zero carbon clothing. But we need to be strongly aware of the impact of reducing our demand for new fashion.

Most of the UK's new clothes are made in Asia. A large fraction of the labour force in developing Asian countries is devoted to the textiles industry. Therefore any reduction in the purchases of clothing by people in Britain will tend to reduce incomes and employment in countries such as Bangladesh, or in Cambodia, where two thirds of all manufactured exports are clothes.

Retailers of clothes are also major employers in the UK high street. One estimate suggests that over 400,000 people work in clothes shops around the UK. Although some fraction of this number may move to businesses that maintain and repair clothes, the move to net zero will inevitably be painful for the retail industry.

We really should pay more for our clothes. And make sure a larger share goes to those who make them.

CHAPTER 7

CONCRETE PROBLEMS

Using less cement and other resoources - and replacing fossil fuels in heavy industry

Globally, steel and cement are each responsible for about 7 per cent of CO_2 emissions. Steel manufacture uses huge amounts of coal, as does making cement, which additionally requires a chemical reaction that emits CO_2 to the atmosphere. Fertiliser production adds another 1 or 2 per cent, largely through the natural gas used to make ammonia. The figures in each case are rising. Even though the developed world's consumption of these resources is close to a plateau, industrialising nations are building huge amounts of new physical infrastructure.

For each of these vital commodities, it's possible to see how production can move towards carbon neutrality.

New steel could be made using hydrogen, while recycling steel is easily possible and already significant. Ammonia can be made from renewable hydrogen. Some of the emissions from making cement could be avoided and, if we systematically increase the use of wood in buildings the world would need far less of this product. In each case, a tax of around $100 for each tonne of CO_2 emitted would be close to making new, low carbon techniques economically viable in the relatively near future.

For all other commodities, the clear imperative is to reduce our consumption as much as possible. The average person uses an average of 33 kilos of the earth's resources every day – almost half the average weight of a human body. About 5 kilos of this consumption consists of fossil fuels, and much of the rest is building materials. Metals make up the bulk of the remainder.

RESOURCES AND PEAK STUFF

Globally, an average of 18 kilos of sand per person are put to use each day, making it by far the most important single constituent of our material use. Although it exists in vast quantities in deserts around the world, most sand is unsuitable for construction and for use in concrete. For these purposes it has to be 'marine sand', dredged from rivers or extracted from beaches. It is in increasingly short supply as the building of new homes, offices and factories accelerates around the globe.

This is typical of the problems the world faces over its use of materials. For centuries we have mined the earth's crust as a free resource. We extract what we want, whether it be gold, lithium, oil, sand or even the natural fertility in the soil, rarely bothering to recycle it, even when we could easily do so. Climate change is just one problem we have created by this reckless attitude. With sand extraction, we now face the widespread destruction of river ecologies, particularly in India and China, and the erosion of beaches that have been carelessly mined.

A few years ago, I wrote a short book about sustainability and put forward a view – now sometimes called the 'peak stuff' hypothesis – that the world's need for physical resources would decline as economies matured. For example, once China had built all the cities it needed it would cease to require huge quantities of steel and concrete (China makes about 50 per cent of the world's tonnage of both commodities). I was both right and wrong. IKEA recently started using the phrase 'peak stuff' to explain the plateauing of its sales in developed markets. But, while Europe is probably using a smaller weight of materials, Asia is more than making up for this slowdown. 'Peak stuff' won't save us from the climate crisis.

CIRCULAR ECONOMY CHALLENGES

Building a working global economy that minimises the need to constantly extract new resources should be the world's target. There is growing talk about 'circular' production processes that reuse existing materials, but most industries have made only token progress towards reducing their footprint. Furthermore, we should be aware that full circularity, in which 100 per cent of recycled materials are put back into use, represents a huge technical challenge. Whether it be the fabrics of a shirt, or the paper of the pages of this book, recycling results in a loss of volume and quality. For example, the cellulose fibres that give us paper are shortened in each reuse, meaning that they can only be recycled about five times.

The problems with metal are different. Broadly speaking, all metals can be recycled without losing significant quality – aluminium being a good example. However, recapturing steel, by far the most important metal in terms of annual tonnage produced, is far from easy. Getting even 85 per cent of the metal back from a demolished building is a very tough target. The world will always need to make new metals.

Plastics can often be recycled, but all suffer serious degradation during the recycling process. Don't assume that the plastic water bottle you have just bought will be processed and reused as a new bottle if you put it in the correct waste bin. The best you can hope for is that it is incorporated in the next polyester fleece that you buy.

Even in a society that has moved substantially towards circularity – a position no country has yet achieved – we will still continue to use new materials extracted from the ground. It is uncomfortable to say this, but buying fewer things, and keeping them for much longer, is the only coherent response to the ecological challenge. The UK's Chief Scientist at the environment ministry made this point forcefully: 'Environmental challenges are not just about emissions. They are about resource consumption.

Emissions are a symptom of rampant resource consumption. If we do not get resource consumption under control, we will not get emissions under control.'

It is worth stressing this point. We tend to assume energy is the challenge in fighting climate breakdown. However, as energy moves to solar and wind, the environmental cost of consuming electricity drops close to zero. The extraction and processing of materials represents a more intractable problem. As well as cutting CO_2 from existing manufacturing processes, we need to reduce the amount of these goods we consume.

The most obvious requirement is to cut our use of plastics. According to one estimate, the manufacture and reprocessing of plastics accounts for about 4 per cent of total global greenhouse gas emissions. Improving low recycling rates is one step. But, particularly in the case of packaging, we need to move towards highly durable containers that last for generations. This is a difficult move but not impossible, and its is already being trialled in France by the Carrefour supermarket chain with the cooperation of such brands as Evian, Coca-Cola and Mars.

IRON AND STEEL

The iron and steel industry creates about 7 per cent of world carbon emissions, mostly from the production of new steel in blast furnaces. Iron ore is shipped, often over very long distances, and then melted at temperatures in excess of 1,000 degrees celsius by burning coal. It is the coal which gives steel its high carbon footprint.

As economies mature, they can make an increasing fraction of their steel from recycled metal. The old steel is collected, perhaps from a building that has been demolished or from a scrapped car, and put into a furnace through which enormous quantities of electricity flow. This melts the old metal and allows impurities to be extracted. Then the liquid steel is allowed to flow out of the furnace and is shaped into slabs for later use. This process uses far less energy than the initial manufacture from iron ore, but even in an economy such as the UK, where overall steel requirements have fallen, recycled steel still provides less than half the total supply. And there are losses in the recycling process. Currently, only about 83 per cent of UK steel is recycled.

This is not to diminish the possibility of more efficient use of steel. Important changes that are needed include extending the life of commercial buildings, which are major users of steel, as well as far greater use of wood in construction. We are beginning to see even multi-storey buildings being made from wood; for example, a seven-floor office block in East London is just being completed. Because they don't use concrete, wood buildings are only a fifth of the weight of conventional equivalents and don't need the same complexity of foundations.

But the world will continue to use new steel and, for this, it must cease burning coal as the source of the heat in the manufacturing process. Most major steel-makers are now working on employing hydrogen made from renewable electricity instead. The only significant obstacle is the extra cost of switching away from coal, which is by far the cheapest fossil fuel for each unit of energy provided. However, a relatively low carbon tax charged on coal consumption would rapidly bring hydrogen to the point where it is cheaper than using coal.

One pioneer project is a joint venture between three Swedish companies that in 2020 will install a new hydrogen steel-making facility in Lulea, close to the iron ore reserves in northern Sweden. Perhaps optimistically, the partnership says that its initial work has shown that 'fossil-free steel will, in future, be able to compete in the market with traditional steel'. The project is small, only intending to make about 1 tonne of steel an hour, but a successful trial will see a full-scale plant established. Sweden is a major steel producer, and conversion to hydrogen fuelling could reduce the country's overall emissions by as much as 10 per cent.

Steel-makers around the world are beginning similar trials, conscious that the use of coal may eventually become impossible, and also that hydrogen made from renewable energy is falling in cost every month. Detailed work from the Energy Transitions Commission, an international business think tank, shows that, even with relatively unfavourable assumptions, the use of low carbon steel need only add about one per cent to the cost of a car. A carbon tax of $100 per tonne would actually make steel forged from hydrogen cheaper than from a traditional blast furnace.

Although the hydrogen route is clear, the investment needed to entirely rebuild a huge and capital-intensive industry should make us wary of how fast the move to low carbon steel can be. New technology can help us, but it needs to be combined with large percentage reductions in the worldwide use of metal. And it either needs a carbon tax or government support.

CAPTURING CO$_2$ FROM CEMENT

A switch to far greater use of wood in construction would also help reduce the emissions from cement manufacture. Strong, 'cross laminated timber' (CLT) can provide a remarkably effective substitute for both concrete and metal structures. A small number of new buildings as high as 50 metres now use CLT. Greater reuse of old materials, avoidance of waste, longer building lives and more careful design can also systematically reduce the amount of use of new cement.

However the cement industry is probably even more difficult to decarbonise than steel and it is also responsible for about 7 per cent of global emissions. Precisely how do they arise?

❱❱ When heated, the raw material calcium carbonate turns into calcium oxide (one of the constituents of cement). CO$_2$ is driven off and enters the atmosphere unless it is captured. This represents about 60 per cent of emissions.

❱❱ Combustion of the fuel required to obtain the high temperatures necessary to make calcium oxide also creates carbon dioxide – adding another 30 per cent.

❱❱ The other operations of the cement plant create about 10 per cent of the greenhouse gases.

Cutting emissions from steel simply requires us to replace coal with hydrogen as the heat source. Unfortunately, this change is insufficient to decarbonise the manufacture of cement because of the extra CO_2 chemically generated by the calcium carbonate.

As an aside, calcium oxide does naturally recapture some CO_2 through a chemical reaction when incorporated in building materials. This is a slow process that can be speeded up by bubbling carbon dioxide through liquid concrete prior to setting. A Canadian company called CarbonCure is the world's leading exponent of this technology and claims that the concrete is actually strengthened by the absorption of CO_2.

Even if such techniques become widespread, however, the world will still need to find a way to capture and sequester the CO_2 arising from the chemical transformation of calcium carbonate. One of the problems is that the gas in the exhaust of a cement plant is not highly concentrated, which makes it expensive to capture. Nevertheless, this may be the only way in which we can neutralise the climate impact of making the calcium oxide for cement. Although many other approaches to making concrete are being tried, some of which avoid CO_2 emissions from chemical changes, none are yet sufficiently advanced to be sure of success.

Standard estimates of the cost of CO_2 capture at cement works are around $110 a tonne. The implication of this number is that a carbon tax of $100 a tonne will not be quite enough to incentivise producers to capture their CO_2. But the figures aren't far apart and carbon capture is likely to become cheaper over time. Most importantly, small-scale CO_2 capture can allow the development of industries that use the carbon dioxide in relatively small quantities in local areas. This will include synthetic fuels that need CO_2 and hydrogen that can be made using surplus electricity from community wind or solar farms.

The second source of CO_2 from cement is easier to decarbonise. At the moment, most cement producers use coal to provide the heat, much as in a steel plant. We can replace this with hydrogen, low carbon biomass or, indeed, simple electricity. The extra cost of this, however, may be high, and the resulting cement may be double the price of today's product once we have also added the price of capturing CO_2 from the chemical reaction.

As the Energy Transitions Commission says: 'Cement is almost certain to be the most difficult and costly sector of the economy to decarbonise'. Rather than despair, I think we need to set our carbon tax at a level that provides a strong incentive both to use low carbon alternatives and to capture CO_2. Because cement is largely used close to its point of production, it is easier for a single country or trading bloc to set its own taxation levels without being undermined by lower cost imports.

And, once again, we need to be much less wasteful in our use of this material.

AGRICULTURAL FERTILISERS

Ammonia is one of the world's most commonly used chemicals, principally in making agricultural fertilisers.

It contains just nitrogen and hydrogen and is produced from natural gas, a process that represents about 1 per cent of global emissions. The good news is that it is becoming possible to make ammonia without significant greenhouse effects, using hydrogen produced from electrolysis. The hydrogen is then combined with nitrogen in the conventional Haber–Bosch process widely used around the world today.

However, the problem is not so much with the production of ammonia but with its breakdown on the farm into nitrous oxide, a powerful greenhouse gas. Ammonia-based (and other artificial) fertilisers also pollute watercourses, rivers and coasts, destroying their ability to sustain fish and many forms of plant life. As the next chapter, on food, makes clear, we need to work out how to sustain our agricultural supply without reliance on huge quantities of ammonia.

Other uses of ammonia, such as in the making of plastics and dyes, do not result in substantial pollution when being used. These products can be made in a much more climate-friendly way by using renewable electricity as the route to hydrogen manufacture.

DOES OUR OWN CONSUMING MATTER?

Our own consuming habits are important. A new computer, for example, might make a measurable difference to a person's carbon footprint for the year of purchase. Apple publishes detailed and well-researched data that shows that a new MacBook Air has a carbon cost of around 176 kilo, including the electricity it uses. That's about 1.5 percent of an individual's average annual footprint in the UK. For comparison, data from leading researcher Mike Berners-Lee suggests that the book you are now reading now has a burden of about 1 kilo, if bought in physical form.

Intelligent companies around the world have succeeded in significantly reducing the greenhouse gases associated with their goods in recent years. For example, careful recycling, the use of renewable electricity and a focus on manufacturing improvements has reduced the footprint of the MacBook Air by almost half in the last two years. When a purchase is necessary, it makes sense to look to buy things from companies such as Apple or Unilever that have made determined efforts to address the climate change impact of what they make. Most large global companies are now publishing properly researched estimates of the carbon footprint of their products.

However the uncomfortable fact is that everything new has a carbon cost. The manufacture of almost any physical item involves some processing that creates greenhouse gases, and nothing is endlessly recyclable (though aluminium comes close). Even as we seek to use renewable energy as the raw material for chemicals such as hydrogen or synthetic fuels, we also need to consume less, make things last longer and purchase goods whose manufacturer is making the strongest efforts to reduce environmental footprints.

CHAPTER 8

PLANT FOOD REVOLUTION

The global climate costs of meat are not sustainable

Around a quarter of world emissions arise from food, which makes this the largest sector after energy. Some authors say the figure is even higher; indeed, one recent report suggested that food emissions may be almost 14 gigatonnes a year – about a third of the global total. For the UK, the percentage is generally noted as somewhat lower, since we import about half our agricultural supplies. But this shows only the complexity and scale of the problem. The flows of foodstuffs make it difficult to assign responsibility for emissions. Other complexities compound the problem; estimates of the climate impact of food depend, for example, on whether emissions resulting from deforestation are assigned to agriculture.

Within food production, by far the most important source of greenhouse gases comes from the production of beef and lamb. This is mainly because cows and sheep emit methane, a powerful global warming gas, from their digestive systems. Significant emissions also come from the use of artificial fertilisers on fields, across almost all agriculture. This has a major impact on climate through the breakdown of these fertilisers into nitrous oxide, another powerful global warming gas.

Routes to get to net zero for agriculture while feeding the 10 billion on the planet in 2050 are hard to find without moving away from meat and switching to very low impact agriculture. Simply moving to organic cultivation will not be enough if fertilisers are imported onto farms in the form of animal manure as the cultivation of those animals resulted in substantial greenhouse gas emissions. Beyond the organic alternative, we need to look at indoor agriculture, radical 'agroecology' and adopting plant-based meat substitutes.

A climate-friendly diet, generally speaking, means eating small quantities of meat and fish combined with large amounts of unprocessed grains, beans and peas, as well as unsaturated oils, such as olive oil, and good amounts of nuts and fruit. Changes in diet and land use are necessary around the world, but getting to net zero emissions will probably be even more challenging than for the energy system.

In the UK, we need to note that intensive agricultural systems are also destroying biodiversity, reducing the carbon in the soil, eroding topsoil and polluting watercourses. In almost every conceivable sense, modern farming is wholly unsustainable. And, whereas people now fully understand the need to decarbonise energy production, very few comprehend the central importance of radical changes to farming.

EAT LESS MEAT

The biggest difference we can make as individuals – and, collectively, to the carbon footprint of food – is to eat less meat. One UK study suggests that simply cutting out meat can reduce the footprint of a typical person's diet by over a third.

By far the most important culprit is the cultivation of cattle for meat (and, to a much lesser extent, for milk). This source alone probably represents 10 per cent or more of global emissions, out of about 15 per cent for all animals grown for food.

The greenhouse gases from cows arise in an array of different ways.

❱❱ Cattle emit large amounts of methane as a consequence of their digestive processes. Methane, partly coming from cows and sheep, is responsible for about a sixth of the global warming of the atmosphere.

❱❱ The manure from cows also produces methane, particularly if it is stored in open slurry tanks on farms.

❱❱ The grasslands on which cattle graze are often artificially fertilised. These fertilisers result in the emission of nitrous oxide, a very powerful climate-changing gas.

》 Cows may eat fresh grass in summer, but generally consume other foods in winter. These grains, grasses and beans will have been grown elsewhere, requiring fertilisers as well as fossil fuel energy to plant, cultivate and harvest.

》 As the human requirement for meat and dairy products rises, the amount of land needed for cows, and to grow the food that they eat, expands. The increase in the global demand for soya beans as animal feed is probably the most important factor driving Brazilian rainforest deforestation.

Beef is one of the most carbon-intensive foods we can eat – although lamb comes close, as does shellfish that has been transported by air. Different researchers give us varying figures, but 1 kilo of beef probably results in the emissions of about 25 kilos of greenhouse gases. By contrast, the same weight of lentils (a good alternative source of protein) results in emissions of less than a kilo. A British family of three, each eating the average amount of beef (about 18 kilos a year each) would see a higher carbon footprint from the meat than from driving a car, or from the electricity used in their house.

There's a second and related argument against beef. About 60 per cent of the world's agricultural land is devoted to the production of cattle, even though beef accounts for only about 2 per cent of all calorie intake. This estimate includes the land given over to cattle pasture but also the area needed to grow the feed for the world's growing herd of animals. We know that reaching net zero around the world will require large-scale reforestation. But if we increase the world's land area devoted to capturing carbon by growing new trees, we will almost certainly have to reduce the acreage given over to cattle production. Continued production of large quantities of beef is simply incompatible with a stable climate.

Are other forms of meat better? The answer is yes – up to a point. Lamb, as noted, has a fairly similar footprint to beef. But pigs and chicken employ a different digestive system to cattle and sheep, and emit much less methane. Largely as a result, the footprint of pork is about half that of beef, while chicken is even lower. There may be benefits from switching from the meat from ruminants (beef and lamb principally) to these alternatives, but all forms of meat, perhaps excluding that produced from genuinely free-range animals, have a climate impact many times that of equivalent protein sources, such as beans or grain.

As carbon footprint expert Mike Berners-Lee notes, part of the reason is that animals are inefficient in converting their food into meat. A cow uses 100 calories of food to make 3 calories of meat. Perhaps even more importantly, from a nutritional standpoint, they give us back far less protein than they actually eat. For every four units of protein fed to animals, mostly made from beans, seeds or grains grown on farms, only one unit becomes available as meat.

Some countries are already successfully reducing red meat consumption. There's been a slight decline in the UK over the last decade, for example. But, to push the level down significantly, we almost certainly need a tax on meat and sustained public service advertising that accurately conveys the products' effect on the global climate, and on human health. This would be unpopular with farmers and food manufacturers, but it is difficult to see any alternative. Government action should also include sensible and sustained support for other forms of agriculture and for the switch from pastoral farming to woodland. (More on this in the next chapter.) Of particular importance, this would help retain jobs in the remote parts of the UK that are largely reliant on sheep farming today.

BETTER DIETS FOR THE PLANET

Modern agriculture, globally and in the UK, pushes many environmental boundaries. It stresses our supplies of fresh water, allows destructive nitrate pollution to reduce the living zones of our seas, reduces biodiversity and encourages our consumption of unhealthy and heavily processed food. It is no accident that the most insightful recent work on how agriculture needs to be changed was published in the *Lancet*, the world's leading medical journal. Poor farming and industrial food manufacture are affecting global health at the same time as weakening most of the planet's critical ecosystems.

The *Lancet* study (freely available online) proposes an ideal diet that delivers 2,500 calories a day. Its fundamental recommendation is that agricultural production 'should focus on a diverse range of nutritious foods from biodiversity-enhancing food production systems rather than an increased volume of a few crops'. Specifically, the *Lancet* diet means getting about a third of calories from grains and just less than a third from legumes (mainly beans and peas). It encourages us to consume nuts and allows for a small amount of meat and fish. For meat, it proposes mostly chicken, and allows only about 7 grammes a day from beef and lamb (this is just over 10 per cent of current UK consumption).

Fruit and vegetables deliver 200 calories a day in the proposed diet, and milk and cheese somewhat less. Oils are more important, providing over 400 calories in this menu, of which by far the largest share comes from unsaturated varieties such as olive and sunflower oils. Palm oil is pushed to a much smaller number, as it is a saturated fat and its production is typically highly destructive to the environment. The central finding from the *Lancet* diet is that the world needs a 'greater than 50 per cent reduction in global consumption of unhealthy foods, such as red meat and sugar, and a greater than 100 per cent increase in consumption of healthy foods, such as nuts, fruits, vegetables, and legumes'.

What characterises the *Lancet* diet is a high degree of reliance on unprocessed foods that have limited climate change impact. But, even were the diet to be universally adopted, we would still see global greenhouse gas emissions of 5 billion tonnes from agriculture in 2050 (that's more than 10 per cent of total current levels from all sources, including agriculture). The authors of the study say that their proposal will have to be accompanied by compensating capture of CO_2, mainly through the restoration of forests and improving the carbon levels of soil. This implies substantial changes in farming practices towards what is called 'agroecology' (of which more below).

The *Lancet* diet is not a particularly radical proposal. It allows continued use of large volumes of fertiliser and does not demand everyone goes vegetarian. But there is no doubt that avoiding meat would help reduce global emissions. And vegan diets, particularly those focusing on unprocessed ingredients rather than manufactured foods, are better still.

If you have decided not to fly, moving to a plant-based diet is possibly the single most important action you can take next. One Oxford researcher says that a vegan eating low carbon foods might be able to cut the carbon footprint of their diet to 27 per cent of a meat eater. A fully vegan diet will typically result in emissions of around half to two thirds of a vegetarian equivalent.

ORGANIC FOODS

Perhaps surprisingly, organic foods aren't necessarily better for the climate, though it depends on the way the food is produced. Organically produced cattle, for example, emit as much methane (in fact, possibly more) than cows that are conventionally farmed, and need to be fed farm-grown foodstuffs in winter. Although organic farmers use no artificial fertilisers, they may need to use far more land to produce a similar amount of meat, burning more fossil fuels to look after the animals.

And this doesn't only apply to meat. One detailed study showed that growing organically produced grapes for wine production had a substantially higher footprint than the conventional equivalent. The organic land on which vines were grown used animal fertilisers, which have a high CO_2 impact, and their cultivation needed more fuel for tractors and other machinery.

That said, the arguments in favour of organic cultivation are not just about their direct impact on greenhouse gas production. The more powerful justifications for cultivation without synthetic fertilisers, pesticides and herbicides revolve around the benefits of working with the natural environment, rather than against it, much as the *Lancet* study proposes. Organic farming allows much more complex ecosystems and biodiversity to develop than those on a large industrial farm. This helps retain carbon. Of course, nature is not fully restored, even on the most carefully cultivated of organic farms, but the soil is not being continuously degraded and the run-off from fertiliser is far less likely to cause pollution damage to local watercourses.

EATING SEASONALLY OR LOCALLY

Food miles matter. However, shipping a product by truck into the UK from Spain probably only adds a few per cent to its carbon footprint. In fact, many studies have shown that locally grown food can be associated with higher emissions if the crop needs to be kept warm in a greenhouse in winter or requires heavy fertilisation compared to the imported alternatives.

It might seem that the climate change arguments for local food, as with organic production, are not especially strong. However, we also need to weigh the benefits of farming systems that help restore the local environment. Food purchased from a farm down the road may well be better for the wider environment; it tends to be unprocessed and thus healthier, and involves less waste than buying manufactured food from a supermarket. These are good reasons for supporting local farms, but probably most important is that small-scale farming is less likely to result in the catastrophic loss of soil carbon that characterises large industrial farms in southern Britain.

One thing we can be sure about is that food products flown by plane have a high carbon footprint and need to be banished from the diet of those wanting to minimise their impact. A kilo of shrimp shipped by air has a carbon footprint at least forty times higher than if it came by ship. In fact, farmed crustaceans brought in by plane are almost as bad for emissions as beef. However, relatively little food is carried by air and the best way of avoiding air miles is not to buy fresh vegetables and fruits that are out of season.

AIMING FOR CARBON NEUTRALITY

In spring 2019, I was lucky enough to visit the Miguel Torres winery, sited in rolling hills close to Barcelona (another enjoyable European train trip). I was there to understand how thoughtful agricultural producers are responding to the climate crisis – and how a winemaker could respond to rising temperatures that will seriously affect yields and reduce quality of wines across the world.

In a hotter climate, grapes will tend to ripen before the complex flavours in a fine wine have had a chance to develop. There are justifiable concerns that the climate in much of Spain will get too hot and dry for premium wines, so the Torres family, one of Spain's biggest wine producers, are planning their response. They've bought large acreages much higher up in the hills, where temperatures are lower, and sourced vine varieties in ancient Catalan vineyards that may be better able to resist higher temperatures. Their new vineyards are planted less intensively, with bigger gaps between rows so that heat isn't trapped, and the leaves are left on the vines over the summer, shielding the precious grapes from excess sun.

Alongside their preparation for the effects of climate change, the family is investing in moving close to carbon neutrality across its chain of production, including its suppliers of bottles, packaging and transportation. It has a fleet of 100 electric and hybrid vehicles. A large biomass boiler burns the spring prunings from each of its vineyards, replacing about 70 per cent of its gas consumption, and solar panels provide a quarter of their electricity. The company is also experimenting with different ways of using the CO_2 that is inevitably produced as sugars turn to alcohol in wine fermentation. All in all, the company will cut its greenhouse gases per bottle by 50 per cent from 2008 to 2030.

That is still not carbon neutrality, of course. So the family has bought 50 square kilometres of pasture in Chile and will reforest it over the next two decades. The plan is that the CO_2 captured by this new forest will counterbalance all the remaining emissions from their wine production and sale. By comparison, current planting of new woodlands across the whole UK average less than 100 square kilometres a year. Such striking commitments to good climate citizenship would probably not happen in a conventional company under pressure to improve quarterly earnings. Family businesses are sometimes strikingly better at responding to environmental challenges.

REWARDING FARMERS FOR BEING BETTER CLIMATE STEWARDS

The Torres family's commitment to carbon neutrality is an example of best practice. But they are not the only pioneers and nor are they alone in recognising the value of looking to the past for adaptation to different crops or varieties that can thrive in a changing climate. In the UK, an interesting proponent of such ideas is John Letts, a seed specialist who grows heritage varieties of grains across smallholdings in the south of England. He sows mixtures of many different seeds no longer used commercially anywhere in the world, many of them last harvested 100 or more years ago.

Letts' diverse heritage plants produce tall stalks, up to two metres high, and their roots are often almost as deep. (If you look at a painting of agricultural life before the twentieth century you will notice that the grain at harvest is often taller than the people wielding the sickles.) Letts doesn't fertilise the fields in which the cereal grains grow, although he does plant a crop to help restore soil quality every third year. Typical yields, he tells me, are about 3.5 tonnes a hectare for his trial fields, less than half the amount gathered from a conventional field. However, since a typical intensive cereal field only produces a wheat or barley crop in one year out of two, the difference in overall productivity is actually quite small.

Why is this relevant to a discussion of food and climate change? When we meet, the first point John Letts makes to me is that his grain does not require use of pesticides, weedkillers or artificial fertilisers, so there will be little or no emissions of nitrous oxide or methane during the production process. He doesn't need to destroy weeds with herbicides, because the grain grows far higher than almost all its competitors and the roots are far deeper, and his grains crowd out competing plants. So there is plenty of light for the plants to grow and plenty of water in the deep soil for them to prosper.

This leads on to the second point. A very deep-rooted crop is far less vulnerable to the summer droughts we have seen in the UK in recent years. John shows me a photograph of one of his fields, full of healthy grain even though the surface of the field is parched and deeply cracked. His yields are unaffected even when neighbouring farms are experiencing the effects of drought.

He can avoid pesticide use because, although individual grain types in his mix may be susceptible to a specific pest, the truly diverse range of grains being grown in the same field is much less vulnerable than a vast plantation of genetically identical wheat, whether grown organically or using conventional techniques. His avoidance of pest control also allows bees and other insects to flourish in the area, helping to ensure that no single creature gains dominance, thus protecting biodiversity.

John Letts works with millers to turn his grain into bread flour. He also supplies specialist food and drink companies, such as a distillery making artisan gin in a disused barn near my home. His heritage flour makes wonderful breads. A sourdough loaf using his grain contains low gluten and a far wider array of micro-nutrients than industrial breads, which have often had gluten added to speed up their manufacture. I suspect that a return to his flours would improve human health as well as the environment.

But this might carry a cost, both financial and possibly in terms of the land required for agriculture. For the last half-century or more, food production has managed to keep up with the global population growth. Nearly a billion people still don't get enough to eat, but the world as a whole manages to grow almost 6,000 calories per person a day, twice as much as we need to be well-fed and nourished. (Some 1,700 calories of this total are used to feed animals, of course; much food is wasted; and 800 calories are used, insanely, for biofuels for transport).

Although global grain yields now show some signs of having reached a plateau, partly because of the effects of the deteriorating climate, the world has grown used to the volumes of food from dwarf high-yielding varieties of wheat and rice and other crops, combined with the application of artificial fertiliser and pesticides. The 'Green Revolution', as it was called, worked miracles for food availability. It also helped prevent vast amounts of deforestation, particularly in Asia, because the area needed to grow sufficient food was far less than it would have been using older types of grain without fertiliser.

John Letts is asking us to think again. Can we combat the effects of climate change by growing older varieties that don't need elaborate care from industrially produced fertilisers and pesticides? Can we do so and still increase the rate of forestation so as to draw down CO_2 from the atmosphere? And how do we make sure that any rise in the price of grains does not impact the living costs of the less well-off around the world?

ROUTES TO LOW CARBON AGRICULTURE

Throughout this book, I have made proposals to move towards net zero emissions within twenty or thirty years. Most of these involve the use of advanced technologies that mitigate human impact. They don't seek to return us to the habits of a century ago. But, if we follow John Letts' route, we will be deliberately turning away from an industrial agriculture that seeks ever higher yields.

Currently, cereal grains provide just under half the world's calories, albeit at significant environmental cost, particularly as a result of fertiliser application. Moving to an agriculture that is genuinely compatible with planetary stability, which Letts' ideas are intended to ensure, we have to avoid meat almost completely. We will need more land for less intensive cereal growing, much as our ancestors farmed 200 years ago.

Providing all human beings with the necessary 2,300–2,500 calories a day and using Letts' approach to cereal growing would involve huge shifts in world agriculture. Even if these changes are eventually seen as essential to climate stability, it will be difficult to marshal sufficient support. Nevertheless, the Letts approach, however eccentric it seems, must be investigated as a way of dealing with the quarter or so of world emissions derived from agriculture, alongside the swing to indoor agriculture and synthetic meat proposed later in this chapter.

What about a halfway house? Noting how destructive large-scale farming is to the environment, many now recommend a solution based around smaller farms practising what is known as 'agroecology'. These farms, often growing fruit and vegetables, are run in a way that both provides food and improves the local ecology. They tend to use renewable energy, often practise fully organic agriculture, and care for the soil and the biodiversity around them. Often they seek to increase the carbon content of their soil, a type of CO_2 capture that we should all be happy to pay for. Many such farms concentrate on selling locally, minimising waste and transport costs. But most of these enterprises struggle to compete against the cheap food in supermarkets.

At the moment, they are not paid much for the services they perform for the common environment. What would happen if our farmers were properly rewarded for not emitting CO_2 or methane, for maintaining the quality of our watercourses or improving biodiversity in ways which benefit other farmers, such as by encouraging pollinating insects?

THE ROLE FOR CARBON TAXATION

One quick check may help us decide how agriculture should develop. The average UK household spends about £60 a week on food, before counting meals out, and consists of an average of 2.4 people. That means a grocery bill of about £25 per person per week, or £1,300 per year. The average UK person has total emissions, including those embodied in things bought from abroad, of around 12 or 13 tonnes. The impact of the agricultural supply chain on greenhouse gas emissions is usually reckoned to be about 25 per cent of this. So let's say 3 tonnes per person per year.

Greenhouse gas emissions should be taxed (as I will propose in Chapter 9), and if we apply a rate of $100 a tonne, at today's exchange rate, we should be taxing our existing sources of food at £250 a year, just for the climate change costs. If we added fees for the other pollution caused by conventional agriculture, then that sum might be doubled. This is a substantial fraction of our yearly food bill.

If farms genuinely following agroecological principles were subsidised for not injuring the environment, their output would be more or less competitive with supermarket foods. As with meat, there is a strong argument for a form of carbon taxation that reflects the huge differences in greenhouse gas emissions between an industrial farm delivering its produce to a food manufacturer and a small producer selling to neighbours.

Whether we need to go as far as John Letts or simply reward farmers for not producing greenhouse gases should be a question under active discussion around the world. The role of food production in speeding environmental decay does not yet have the political importance of burning fossil fuels. This needs to change.

MOVING PRODUCTION INDOORS

A hundred feet below Clapham in south west London lies the world's first underground farm. In tunnels that functioned as air raid shelters in the Second World War, the Growing Underground team cultivates a variety of micro-leaves and herbs. Ranging from red amaranth leaves to pea shoots and wild rocket, the produce is grown in trays that are stacked from the floor to the ceiling. Using powerful LEDs to provide high levels of reddish growing light and nutrient-filled water flowing through the trays, the plants can be grown quickly, without weeding or pesticides. 'Hydroponic' farms like this are, of course, completely unaffected by the weather on the ground above them, and produce can be grown throughout the year. The salad and herbs are harvested just weeks after sowing and sold through the large supermarket chains.

A similar approach, on a far larger-scale approach, is employed by the Jones Food warehouse in Scunthorpe, which grows a mixture of herbs and green leaves. It ships more than 400 tonnes of produce each year to wholesale customers, including the UK's largest sandwich maker. The whole 'factory' (this time above ground) is automated, using a robot to move the trays up to the highest level. Only eight people are employed to actually produce the food. The online grocery retailer Ocado bought a large stake in the company in 2019, announcing plans for a partnership to open another ten indoor farms within five years. The Jones Food CEO said that the aim is to put warehouses for growing food next to Ocado's warehouses, hoping in some cases to get the fresh leaves to customers within an hour of being picked.

The central benefit of this new type of farming is that growing vegetables and fruits inside a building, usually in trays that are stacked vertically as if in a warehouse, improves yields compared to a field of the same size. It is claimed that productivity per square metre is a staggering 350 times the level of outdoor agriculture. Other advantages include huge savings in water and fertiliser use. One Japanese indoor farm states that it uses only about a tenth of a litre of water for each head of lettuce, less than 1 per cent of what is needed in fields. As climate change increases the risk of serious drought in agricultural areas, this advantage will become increasingly important.

The list of benefits continues. Today, most indoor farms are located in cities and so the produce minimises transport emissions. 80 Acres, a vertical farm operator in Ohio, claims that the growing cycle allows seventeen harvests of salad per year, compared to two in conventional outdoor agriculture. Plants also absorb CO_2, so we can hope that captured greenhouse gas will be piped into indoor farms to improve growth rates. (One of the uses for a Climeworks CO_2 capture machine – mentioned in Chapter 11 – is to produce carbon dioxide for a Swiss greenhouse).

Although indoor agriculture uses gargantuan amounts of electricity to produce the light that the plants need for healthy growth and then to remove the heat from the LEDs, in other respects it is environmentally benign. No pesticides are necessary, there's no run-off of fertilisers into rivers and the risks of bacterial contamination are much reduced. The proponents of indoor farming say that it will eventually mean that the world can let large amounts of agricultural land revert to forest.

Does the increased electricity use matter? Possibly less than we might think. Many indoor farms use renewable electricity, knowing that it is part of their appeal to customers that their product is truly sustainable. Electricity is, of course, expensive compared to the sunlight that conventional growers rely upon. However, as low carbon electricity becomes ever cheaper and more abundant, such concerns will drop away. Many farms will eventually be powered by solar panels sat on the roof of the building or on unused land nearby, combined with batteries to ensure the lights can be turned on every day.

Indoor farming is growing rapidly around the world. The expansion is often funded by investors aware of the likelihood of environmental change substantially reducing outdoor yields of food. Most of this first generation of indoor farms concentrates on salads, leafy vegetables and herbs such as basil or coriander. These are the easiest products to grow indoors, but some companies are experimenting with other vegetables. 80 Acres grows strawberries, as does the French company Agricool, which uses highly insulated shipping containers. Tomatoes and cucumbers are also possible crops.

As someone who grows vertically using hydroponic nutrition, albeit inside a little greenhouse on an allotment, I can testify that the vegetables and herbs grow very quickly, almost invariably without disease, and with high standards of taste and appearance. (I am particularly proud of my pesto, made with luscious thick leaves of basil that only take a few weeks to grow.) It enables me to get more productivity, and flavour, out of the tiny plot of land that I rent.

However successful indoor or hydroponic agriculture becomes, it is unlikely to provide a significant fraction of the world's calories. The *Lancet* study referred to earlier sees salad vegetables as constituting only about 3 per cent of all calories in the ideal diet. And, although indoor growing will save land, the overall amount will not be huge. (Only about 2 per cent of the UK's agricultural land is given over to horticulture today, although this is chiefly a consequence of buying 95 per cent of our fruit and at least half our vegetables from overseas.)

Nevertheless, one definite benefit of any move to indoor cultivation of vegetables and fruits would be an increase in local employment. Indoor farms are not large employers, particularly if they use advanced robotic techniques, but they will add year-round jobs across the UK, producing foods that are generally grown for us today in the rest of Europe. And these jobs can be distributed across all regions of the UK. They can be part of the Green New Deal that we need to achieve.

PLANT-BASED 'MEAT'

A potentially more significant development is the move to create food that looks, feels and tastes like meat but does not come from animals. Most of these new food-stuffs are manufactured from plant matter.

The best known is the remarkably meat-like burger from Impossible Foods Inc. Impossible makes its product so that it looks and tastes exactly like conventional fast-food restaurant meat. The secret to its flavour lies in imitating meat by using a substance ultimately derived from the roots of soya beans. The gene for this product has been inserted into a fast-growing yeast for speed of production. The original material it uses is called soy leghaemoglobin (also known as 'heme') and contains iron, which Impossible says is critical to the flavour of beef.

The raw burger's other ingredients are water, coconut and sunflower oils, and a soya concentrate. When cooked, an Impossible burger is intended to 'bleed' a reddish liquid similar to the blood in meat. In all respects, the food mimics the original beefburger. Today, a Burger King Impossible burger sells for about $1 more than the equivalent item made from real meat. This premium is likely to disappear as soon as production scales up, partly because the raw cost of the Impossible Foods ingredients is so much lower than real meat.

As you might expect, the Impossible Food's burger has a carbon footprint that is radically lower than meat. The company claims an 87 per cent saving, and similar improvements in water and land use. Yes, the company still has to buy soya products, which require land to grow, but any plant matter is vastly more efficient than cows in terms of production per square kilometre.

Beyond Meat is another US contender in the competition to replace meat with plant matter. It produces products that sell in supermarkets, such as the largest Tesco stores, and is trialling product with McDonald's. Fully vegan, the burger matches the Impossible Foods product in taste tests. The company also makes 'sausages' and 'meatballs' that are marketed in the same fridges as real meat in stores. The taste and appearance of meat is created through coconut oil and the use of deep red beetroots, while the main ingredients are pea, mung bean and rice, all easy to produce in a reasonably environment-friendly way. In addition, the products can justifiably claim to be better for human health than real meat. The company, newly listed with enormous success on the New York stock market, claims similar greenhouse gas, water and land use benefits to the numbers provided by Impossible Foods. There's little doubt that the interest of big investors derives from the trillion dollar size of the potential market for alternatives that can imitate animal-based foods. Beef substitutes are only the start.

These meat equivalents seem to appeal and can even be confused with the real thing in blind tastings – and within a few years they should be far cheaper than the animal alternative. Many people, of course, will remain loyal to conventional products, but worldwide we are seeing a growing realisation that animal cultivation is bad for the planet, bad for human health, as well as a crime against sentient creatures.

What would be the consequences for local employment and patterns of land use if there was a full conversion to growing plants as raw materials for meat substitutes, instead of keeping animals? Would it help the UK? The labour needed to grow a tonne of peas or soya is probably less than the equivalent for beef or sheep. And these crops are not well suited to being grown on small farms, so we can expect the extra production needed to stay in the areas where such crops are currently grown. So we probably won't see much of a local benefit from the near-inevitable growth of 'meat' made from plants. Note, however, that many of the ingredients for the new meat substitutes, such as peas, are well suited to UK growing conditions.

Full-scale industrial production of meat substitutes could, however, have climate benefits almost as large as the decarbonisation of electricity production.

FARMLAND INTO WOODLAND

Just under 30 per cent of the UK is pastureland, used by farmers for animals. As meat consumption falls, less and less of this will be needed. We will also require less land for growing crops, since so much production today goes to feed animals.

Conversion to woodland could provide an economic future for the UK's pastureland and its farmers, as well as drawing down carbon from the air. This topic will be more fully covered in the next chapter, but the key numbers are these: if we converted 1 per cent of the UK's pastureland each year until 2050, we would have created 700 square kilometres of forest a year, which would sequester the best part of 1 tonne of CO_2 per person per year. That is more or less equal to the food chain emissions of a vegan – which illustrates that farming and food production can become carbon neutral – and just how much we need to change our eating habits if we are to achieve this.

CHAPTER 9

REFORESTING BRITAIN

Using forests and woodland
to suck CO2 from the air

UK carbon emissions are partially counterbalanced by the takeup of carbon dioxide by trees. Currently, a net 24 million tonnes of CO_2, equivalent to about 5 per cent of current domestic greenhouse gases, are extracted by photosynthesis in the UK's woodlands. Getting to net zero will be helped if we increase the 'negative emissions' from trees. However, unless the UK manages its forests better, and increases its coverage, the rate of sequestration by woodland will tend to fall over the next decades. Many of our forests are now old, and the trees are ceasing to grow rapidly, reducing the capture of CO_2.

The proposal in this chapter is to turn the UK's poor quality grazing lands, which cover about a sixth of the country, over to forestry. This is a radical suggestion, involving the conversion of 40–50,000 square kilometres of land. However the Forestry Commission says the land is available and in moving to 30 per cent woodland we would be doing no more than matching the land use of other large European countries, most of which are about one third forested.

CARBON CAPTURE FROM FORESTRY

As we all learn at school, trees takes carbon dioxide from the air and use it to grow. Photosynthesis is therefore a natural form of carbon capture. The carbon is stored in the roots, trunk, branches and leaves of the tree for the whole of its life. This might be for hundreds of years in the case of large broadleaved trees, such as oaks; or less in the case of conifers, including the widely-planted Sitka spruce.

The UK is one of the least forested countries of Europe. Scotland and Wales have greater coverage, but England is only about 10 per cent forest. By contrast, Germany is 32 per cent woodland and Italy slightly more. The Nordic countries are two thirds forest. Our land use is very different in other ways to most of our neighbours; 17 per cent of our land is given over to rough grazing, principally for sheep, and it is primarily this area that I suggest should be given back to forestry.

The UK has managed to marginally increase its woodland cover in recent decades, but the rate of planting of new trees has fallen to less than a third of its level thirty years ago. Without a major initiative, the amount of carbon being turned into wood will fall in the years to come, because our trees are passing their point of maximum growth. Precise numbers are hard to come by, but at the moment UK trees probably take in about 20–25 million tonnes of net CO_2 a year, equivalent to offsetting nearly 5 per cent of national emissions.

Over the past twenty years, UK governments have toyed with the idea of large scale reforestation and proposed a substantial increase in tree planting. But we have seen no increase in actual planting rates for some time. The National Forest, which runs across parts of the Midlands, was started a quarter of a century ago and has now planted about 9 million trees. The effects are highly beneficial in many ways but the scheme outlined in this chapter would plant that number every six days for the next thirty years. Is this too ambitious? Possibly, but the people of Ethiopia planted over 350 million trees in a single day in July 2019 as a response to the deforestation of their country.

A NATIONAL TREE PLANTING PLAN

The proposal in this chapter is that we largely replace sheep farming on the UK's poorest soils with the growing of wood. Farmers keeping sheep have been subsidised at more than £200 a hectare (about £80 an acre) under the EU Common Agricultural Policy (and the government has promised to maintain this figure post-Brexit). The animals they grow are worth just over £1.2bn a year at market price. But, at the same time, the UK is importing wood products costing about £8bn that we could have grown in roughly the same amount of space. In other words, in choosing to support sheep farming we are backing an industry which delivers less than a sixth of the total value that could be provided by wood products. It would be good for farmers and good for the country if we replaced sheep-growing areas with woodland.

Of course, we need to keep in mind that sheep farming is deeply embedded in many of Britain's rural areas, and the transition to a local economy based on wood, rather than livestock, will be intensely painful for many. However, the uncomfortable truth is that sheep farming is financially costly and environmentally unsustainable due to its substantial methane emissions, its inhibiting effects on biodiversity and its increase in vulnerability to flash flooding. We need to change this and to find alternative incomes for sheep farmers.

Is it conceivable that we could reforest this much of the UK? Increasing woodland from 13 per cent to 30 per cent of the UK's land area by 2050 will mean that about 1,400 square kilometres will need to be planted annually. That rate is almost three times the maximum ever achieved in the UK in the early 1970s. It is a very demanding target. However, other European countries have succeeded in adding woodlands at a not dissimilar rate. France added 2.6m million hectares in twenty-five years and is now about 32 per cent wooded (exceeding our proposed UK figure), and now has about six times as much forested land as the UK. Italy has achieved similar reforestation rates in recent decades and today has three times as much wooded land today as we do.

But can we achieve this target while also meeting demands for land for our expansion plans for solar PV and onshore wind farms and 100 per cent renewable energy? This answer is fairly straightforward. Appropriate areas for trees are not usually the best for solar PV. Forestry development will be concentrated in areas of upland in Wales, northern England and Scotland, where sunlight levels are not optimal. Most PV will go in the south-east and south of England. And, as half the UK's land area is currently given over either to rough grazing (17 per cent) or permanent grassland (31 per cent), merely by substantially reducing animal cultivation we will have enough space to build the basis of a net zero society.

Improved tree cover also has other significant benefits. It reduces the run-off from rainstorms and significantly cuts the risk of serious floods. Trees help keep temperatures down, particularly in urban areas because water evaporates from their surfaces, cooling the air. Forests may also aid cloud formation, which is helpful in areas of potential drought.

Of course, woodlands also offer wonderful walks and other recreation, and the potential for a diverse range of animal life, particularly when compared with the waste-lands created by upland sheep. In a recent Forestry Com-mission survey, 88 per cent of British residents said that 'a lot more trees should be planted'. We will need to use most of our upland grassland for this, but should be able to retain the iconic vistas of our National Parks.

WHAT TO DO WITH ALL THE WOOD?

As noted, the UK is a major importer of wood and wood products. Imports totalled £8.0bn in 2018, while exports were worth only £1.8bn. This means that the country brings in more wood than any other country in the world, with the exception of China. As the Royal Forestry Society states: 'The UK utilises 50 million cubic meters of timber annually, 10.6 million cubic metres of which is grown in the UK.'

A plan to increase woodland cover to 30 per cent of total land would allow the UK to become approximately self-sufficient, but would probably need tax help to encourage full management of the valuable wooded areas. There are many ways to do this that would cost far less than the current sheep subsidy programme. Logically it makes sense to pay wood farmers for the CO_2 they capture using the carbon tax proposed in Chapter 10.

One of the main benefits of increasing forestry is to help grow incomes, as well as sufficiency in energy and wood products in some of the UK's remoter and generally less prosperous regions, where tree planting will be concentrated.

And this is how the wood could be used:

>> Wood pellets. The UK imports about 7 million cubic metres of wood pellets each year, mostly for power stations such as the enormous unit at Drax in North Yorkshire. At present the pellets are imported from the US and Canada and only a small amount is grown in the UK. In other countries, small biomass

power stations burn pellets produced locally to make electricity, often directing the waste heat to district heating systems for nearby homes and businesses ('combined heat and power').

)) Firewood. Alternatively, wood chips can be used for heating large buildings such as hotels or hospitals. Once again, the fuel can be produced and used locally, increasing employment.

)) Sawn wood, wood panels and other products for construction. Wood that is used for long-term construction stores carbon for centuries. New housing in the UK should be principally made from locally sourced wood, as it is in many other countries.

)) Pulp and paper. This is more difficult, because paper mills have to be constructed at large scale to make sense financially. We would be unlikely to see a rapid rebuilding of the UK pulp industry in response to reforestation. But, over time, the far greater availability of reliable, low cost wood supplies could support a paper products industry.

WHAT SPECIES SHOULD BE PLANTED?

Conifers produce a greater weight of wood per year than broadleaved varieties, so a plantation of Sitka spruce may well be the route to the maximum output per hectare. Nevertheless, it makes good sense to plant a diverse selection of tree types across the country, both for reasons of biodiversity but also to protect against new diseases and pests.

Some tree species that existed in the UK prior to the last Ice Age disappeared here but still grow well in continental Europe. They should be brought back into cultivation. This will improve resilience to climate change, pests, and perhaps also to the risks of fire.

In addition, we could increase the land area given over to trees and plants that can be repeatedly coppiced to generate a continuous stream of fuel for burning. This includes a very tall grass called Miscanthus and the hazel, a tree that is well suited to lowland UK climates.

WHAT ELSE SHOULD WE BE DOING?

Apart from rapid reforestation, there are a number of other steps the UK could take to improve the rate of carbon capture in soils, plants and trees, at the same time protecting the country's biodiversity.

》 Restoring peatlands. Properly managed peatlands store large amounts of carbon permanently and cover about 12 per cent of the UK. However, our areas of peat are not being properly looked after and today are substantial net emitters of CO_2. Where possible, peat needs to be kept permanently wet. Allowing peatland to dry out causes permanent loss of carbon. Some of this land is used today for arable cultivation, so we will need to resolve conflicts between agricultural production and emissions reduction. Wet peatlands can support some species of trees, although most are intolerant of the conditions.

》 Improving hedgerows. The UK has lost at least 400 square kilometres of hedgerow in the last forty years. Replacing this would improve biodiversity and reduce soil loss from high winds, as well as storing CO_2.

❯❯ Managing existing woodlands for biodiversity. Only about half of English (not British) woodlands are actively managed today. They are often overcrowded and are poor habitats for birds and other animals, as well as gradually losing carbon as the wood rots. The forest floor is too dark to sustain any plant life. Re-establishing careful thinning of these woods and active use of the surplus wood will have wider benefits, in addition to increasing CO_2 storage.

ARGUMENTS AGAINST ACTIVE MANAGEMENT OF WOODLAND

Many environmentalists support strong measures to increase the land area given over to trees. However, they often believe it is wrong to manage these forests and to exploit them for wood and particularly oppose the burning of wood products for energy purposes. Their argument is that this reverses the capture of CO_2 and means that increasing the size of the forested area has no positive effect on emissions.

Others reject these views, saying that active management of forests – for example, taking out excess trees – adds substantially to the rate of growth of the remaining wood, and thus enhances net capture of CO_2, as well as providing a far better environment for species diversity. They also maintain that cutting trees when they are fully grown, and burning the resulting timber, doesn't necessarily mean that forests don't hold on to carbon. CO_2 continues to be stored in a managed forest, provided that felled trees are always replaced with new saplings, ready to grow. The half-century cycle of planting and felling should also keep improving the carbon levels contained in the soil in which the trees live.

This is an important debate. My view is that, as long as we ensure that all trees that are felled are replaced by successful new plantings, we can be confident that expansion of the wooded area will be of real benefit to the climate, as well as to the economies of the UK's remoter regions.

CHAPTER 10

CARBON TAXATION

The economist's answer to the climate crisis

Almost everything that we use or consume results in the emissions of greenhouse gases, whether it is a burger in a restaurant, taking a flight or buying a car. If each item is levied according to the climate change damage it does to the planet, we can use carbon taxation to discourage us from consuming goods or services that have a heavy environmental impact. But, better than that: if we raise the carbon tax to high enough levels, we will encourage suppliers to offer us cheaper alternatives that don't result in substantial emissions.

The easiest way to apply the tax is usually thought to be a levy on fossil fuels, which then feeds through to the products sold to us. Perhaps surprisingly, such a tax is actually widely supported by large companies, even those who produce oil and gas, because it would weigh equally on all participants in an industry and encourage innovation.

But any form of carbon taxation that affects those on lower incomes is unlikely to be politically palatable. We know this from resistance to petrol taxes in France, which helped create the Gilets Jaunes movement, as well as fuel tax protests in the UK in 2000 and 2007. These protests all arose because less prosperous people felt they were paying disproportionate tax. However, there is no reason why a carbon tax should not be used to equalise incomes, rather than penalising the poor. My proposal is that the money raised by carbon taxation should be redistributed as lump sums to all UK citizens, ensuring that the tax provides a net benefit to the less prosperous.

CARBON TAX AND THE MARKET

Those who believe that free markets and conventional capitalism can save the planet will sometimes assert that carbon taxation is the only policy required. With some justification, they argue that imposing a tax on the emissions of greenhouse gases is by far the simplest and most efficient way of reducing the use of fossil fuels, as well as improving world agriculture so that it emits less methane and nitrous oxide (the second and third most important greenhouse gases).

By applying a standard carbon price across all parts of the economy, we could, in theory, systematically disadvantage all activities that cause global warming. A litre of aviation fuel, for example, produces about 2.5 kilos of CO_2. The figure is approximately the same for petrol, while natural gas produces about half a kilo of CO_2 for every kilowatt hour that is generated. The production of beef, which is one of the highest carbon foods, could be loaded with a tax corresponding to the 5 kilos of greenhouse gases that arise as a 200-gramme burger is produced.

The logic is impeccably clear. When a household or a business burns a lump of coal, they gain the heat from the fuel. But, at the same time, they add CO_2 to our shared atmosphere, imposing a cost on the rest of us. There is a striking imbalance between the benefits of the heat and the costs borne by the rest of the world's population: for every kilowatt hour of heat produced by the coal, the wider atmosphere heats up, over the life of the CO_2, by about 100,000 times as much. In other words, what economists call the 'externalities' of burning fossil fuels are vastly greater than the direct benefits.

We all understand the principle that the 'polluter pays'. So, having put some CO_2 into the air, how much should we be taxed to compensate others for the pollution we have caused? According to scientists researching this, the future damage caused by emitting an extra tonne of carbon dioxide will be about $400. Thus, to fully recompense the rest of us, there should be a charge of this amount.

Most economists suggest a more modest level of carbon tax of $50–100 per tonne, a figure that is designed to create a level playing field, and thus to incentivise business and consumers to change to low carbon or zero carbon products. A tax at this level would help all of us – businesses, governments and individuals – see the impact of our choices of what to produce and to buy.

More importantly, taxing greenhouse gases at $100 a tonne would dramatically change the economics of fossil fuel use. To take one example mentioned earlier, it would make the use of renewable hydrogen competitive with the use of coal for manufacturing steel and cement.

The combustion of a tonne of coal puts about 3 tonnes of CO_2 into the atmosphere, meaning that a $100 carbon tax should result in a levy of $300 for its use. Compare that figure to the current (late 2019) price of coal on world markets of around $70 a tonne. A $100 carbon tax would therefore more than quadruple the price of using coal. The percentage impact is much less severe on oil and gas ,as the table below shows. For comparison, I have also included an approximate estimate of the impact on beef prices.

IMPACT OF $100 CARBON TAX ON PRICE OF FOSSIL FUELS	
Coal	+ 430 per cent
Gas	+ 110 per cent
Oil	+ 60 per cent
Beef	+ 55 per cent

(June 2019 prices, oil at £50 a barrel, gas at 40p a therm, coal at £70 a tonne, Tesco beef patties at £4.50 a kg)

Despite enthusiastic support from economists, the use of carbon taxation (or carbon pricing, as it is sometimes known) is barely discussed by environmental campaigners, who tend to prefer action plans which directly target specific causes of pollution. This is a mistake as we need both to encourage new low carbon technologies and to ensure that polluters are financially encouraged to abandon the dirtiest activities. Taxation is at least as effective as regulation.

INTERNATIONAL TRIALS

Progress towards effective carbon pricing has been slow, although certain countries use it very effectively. Sweden, for example, has a carbon tax that has both driven down emissions and been broadly popular. This scheme began, impressively, in 1991 and the level of the tax has risen so that the country now charges about £115 (or over $140) per tonne of CO_2 emitted. During this period, fossil fuel use for domestic heating has fallen by 85 per cent and most households have switched to heat piped through community networks, or the use of low carbon fuels such as wood pellets. In fact the tax has been so successful at driving out fossil fuels from domestic heating that it now raises little revenue. Tax on transport fuels, however, brings in sufficient funds that Sweden claims it has been able to raise the income tax threshold as a result. This tax has therefore tended to reduce inequality of post-tax incomes, one of the objectives of our New Deal for Climate.

A less successful model is offered by the Canadian province of Alberta, which introduced a carbon tax in 2017, focused on motor fuels and heating in buildings. Gasoline was taxed at about 5 pence per litre, equivalent to a rate of about £75/$90 a tonne of CO_2 emitted. However, the province abandoned the scheme in 2019 as a result of political opposition. This happened even though individual households received flat-rate payments each quarter, sharing out all the revenue the government expected to receive from the tax. Most taxpayers were actually net beneficiaries from the scheme, but nevertheless its unpopularity saw it removed.

The UK already operates what has probably has been the most effective carbon tax in the world, an underappreciated achievement and one which has helped reduce the country's emissions from electricity generation faster than almost any other developed economy. In 2012, just before the country introduced a new levy on carbon emissions, coal was generating 40 per cent of all electricity and the UK was emitting 125 million tonnes of CO_2 from burning coal in power stations. This fell over 80 per cent to 22 million tonnes in 2016 as an £18 per tonne levy caused the fuel to become uncompetitive with natural gas. The decline of coal has continued and weeks now pass without a lump of coal being burned to generate electricity.

This very simple (and quite low) tax has pushed highly polluting coal off the UK network, almost certainly for ever. By 2021, almost no power stations using this fuel will remain open. Here on our doorstep we have seen the power of taxation to change the direction of an industry within a just a few years.

But can this principle be extended in order to change our energy system, and indeed every aspect of consuming and production, more profoundly? And could it ever be applied on a global basis? It is hard to imagine all countries agreeing. Nevertheless, a tax applied at international borders on exports from countries not levying a carbon fee is possible, and is already being investigated by the European Union.

INCENTIVISING OIL AND GAS COMPANIES

A high enough carbon price would encourage the development of a low carbon society if it could affect all use of fuels, not just electricity generation, and so impact anything that used fossil energy. And, as noted, many major oil and gas companies support the idea. BP, for example, says 'a well-designed price on carbon is the most efficient way to reduce greenhouse gas emissions'.

Many of us will instinctively distrust statements like this from large fossil fuel companies which are implicated in the emerging disaster of climate breakdown. Cynically, perhaps, we think that they will argue for a carbon tax, but then try to persuade us that it should be kept at levels that do not really impact their businesses.

I'm more optimistic. I think that senior managers and other employees at fossil fuel giants are increasingly aware that their companies are becoming social pariahs and, as individuals, they want to make the switch to a cleaner future. But the pressure from shareholders to keep up the level of short-term dividend payments is enormous and a move to a more sustainable future will undoubtedly depress short-term profits.

A sufficiently high carbon tax would release them from this trap, because they will be able to show shareholders that the financially rational course of action is to pursue low carbon opportunities. A high carbon tax could, for example, persuade companies to invest in synthetic fuels made from hydrogen and captured CO_2 (see Chapter 1) rather than opening new oil or gas fields.

As another illustration, imagine an oil company deciding whether to spend $5bn exploiting a new offshore oilfield or, as an alternative, putting the cash into wind farms in the North Sea. A meaningful carbon tax of $100 per tonne of CO_2 that was applied globally would raise oil prices by about $40 a barrel or more, meaning that a field already expensive to develop would make no financial sense. Meantime, the rise in oil prices will increase the speed of a global switch to electric cars, increasing the demand for electricity and thus probably improving the prospects for offshore wind.

WHY DON'T WE HAVE CARBON TAXES?

Why, then, isn't implementation of a carbon tax more widespread, or a significant demand for campaigners? And why have carbon taxes proved so controversial and difficult to raise to levels that affect decision-taking?

One answer is that a carbon tax means rising prices. Its very intent is to raise the price of things that emit greenhouse gases in manufacturing or use. So a tax will, for example, raise the price of petrol and make driving more expensive. Nobody likes paying more for necessities. In the case of domestic electricity and gas, poorer families have to spend a larger part of their income than the well-off. While the UK's richest 10 per cent of households use 3 per cent of their expenditure for utilities, the poorest 10 per cent spend 8 per cent. Unless neutralised by a tax cut elsewhere, a carbon levy on gas would make household finances even more difficult to manage for the least well-off.

Taxation is a blunt weapon and fairness is hard to legislate. A rise in the price of electricity affects schools and hospitals as much as wasteful households who use the tumble dryer too much. A carbon tax raises the price of fuel for everybody, even though wealthy city dwellers may barely use a car while rural inhabitants often have no choice but to drive a car to get to work. And, of course, almost all forms of taxation are national, but carbon taxation will only work properly if it is applied globally. Otherwise, a tonne of steel made in a country which does not use a carbon price will be cheaper than metal from a blast furnace located where there is a tax. One way around this is to impose equivalent customs duties on goods imported from countries that refuse to implement carbon pricing, but it is problematic to check on the ultimate source of most commodities. Another would be to levy a carbon tax only on activities not facing international competition, such as the supply of gas to homes. It wouldn't be perfect, but it would help to reduce emissions.

Lastly, a carbon tax is not popular among politicians, because it does not encourage their favoured new technologies or advance their constituents' businesses. It is easier to argue for a subsidy for solar panels to please an installation company in their home town than it is to vote for a carbon tax that might have the same effect, but through a much less visible mechanism.

Nevertheless, a carbon tax has many strengths, and we should push our leaders to implement it, alongside well-considered schemes to compensate the less well-off. This might include a rural subsidy for buying electric cars or a better-funded public transport network. The route to a zero carbon world will never be taken if it makes life more difficult for large numbers of ordinary families.

FREQUENT FLYER DEBITS

A good place to start a carbon taxation scheme might be with air travel. Flying is concentrated among the wealthy. The richest 10 per cent of UK households spend almost £1,000 a year on tickets, while the poorest 10 per cent spend less than £100. In fact, as noted earlier, just 15 per cent of the population (almost all with high incomes) make 70 per cent of all flights from the UK, while almost half of all adults in the UK don't fly at all in a typical year.

One suggestion to deal with this inequality is to hand everybody in the country a carbon voucher for one return air flight to Europe each year and allow the people that don't travel the right to sell it on an online exchange. Those who fly a lot would need to buy the vouchers from those who don't. The effect would be to redistribute income from rich to poor. Frequent flyers would face a significant charge for their trips, but most will be able to pay the price. Those who didn't travel by air would see their income increase. The attractiveness of this scheme is that it would be transparently fair. But it would also be complex to administer and manage.

A much simpler route would be simply to increase air passenger duty each year, though this wouldn't provide such a visible reward to those who don't fly. This duty is already in place and is a reasonable mechanism for allowing countries to gradually tighten the restrictions on the total number of flights. The UK applies it as part of its regular taxation (current rates are £13 per person on short-haul economy flights to most of Europe, and at least £78 on long-haul flights), but not explicitly as a carbon tax. Its proceeds could be linked directly to carbon capture and storage, so that the CO_2 generated by any journey is actually removed from the atmosphere (more on which in the next chapter).

REWARDS OF A CARBON TAX

Establishing a direct link between carbon taxation and the cost of removing CO_2 from the atmosphere has clear appeal. If our actions produce 1 tonne of carbon dioxide, we should be willing to pay the full price for its removal. But politicians might prefer a different scheme that instead gave each citizen a yearly dividend, say £1,000, as a rebate of carbon taxation. This would mean that less-well-off people, who will generally have paid less than this amount in increased prices, will be net beneficiaries.

We would need to have effective plans in place for compensating those unfairly affected, partiucarly in rural areas. But I think the arguments for a carbon tax at this level are overwhelmingly strong. Why?

First, because it will move fossil fuel companies from being opponents of the energy transition to being active participants. It's time to exploit the willingness of oil and gas companies to be forced to make dramatic changes. A financial analyst said to me that the fossil fuel companies are 'betting against humanity'. A carbon tax changes the odds, and helps them start betting on the right side.

Second, because a flat carbon tax across fuels and other sources of greenhouse gases doesn't force us to make investment choices that may turn out to be flawed or too expensive. We can allow businesses to experiment – sometimes losing money, but eventually finding the best route. Properly incentivised by a high carbon tax, business can be profoundly helpful, rather than obstructive, short-sighted and dishonest. A carbon tax can be allied with government support for research and development targeted at new renewable technologies.

And lastly – and perhaps most significantly – because that tax figure of £1000 a tonne almost certainly exceeds the eventual cost of extracting CO_2 directly from the air (as we'll see in the following chapter). So any entity still producing CO_2 will be able to save money by genuinely 'offsetting' its emissions, paying for an equivalent amount of CO_2 to be removed from the air and then stored.

CHAPTER 11

DIRECT AIR CAPTURE OF CO$_2$

A vital technology for reducing carbon dioxide levels

As we've seen, some areas of the economy will prove hard to move to full carbon neutrality. Food production will continue to have substantial emissions. So, too, may cement manufacture, and probably clothing. Aviation emissions are a tough problem to solve. To some extent, we can counterbalance these remaining emissions by reforestation. But planting trees may not be enough. We will probably also need to remove CO$_2$ directly from the air. And, if we don't move rapidly enough to cut emissions, we will need to make up for lost time. We may literally have to remove years of emissions.

The technology to do this is relatively simple, but it is energy-intensive. We will need cheap, abundant renewable electricity, and inexpensive heat, to pull the cost down to below \$100 per tonne of CO_2. Most of this captured CO_2 will need to be sequestered underground, which is perhaps as great a challenge as its capture, though we can use some of the CO_2, as well as hydrogen, for making synthetic fuels, such as low carbon aviation kerosene.

DO WE NEED AIR CAPTURE OF CO_2?

Even if we transfer quickly to an energy system based exclusively on renewables, the world will still produce greenhouse gas emissions – and probably billions of tonnes a year. Conventional carbon capture technologies won't help us very much, because they will only work with large fixed sources of emissions, such as power stations.

Moreover, scientists are suggesting with increasing frequency that getting to net zero by 2050, the target that is increasingly being adopted around the world, is possibly not sufficient to keep the eventual temperature rise below 2 degrees. Temperatures are driven by the accumulated weight of greenhouse gases in our atmosphere. So, the longer we wait to start driving down emissions, the higher temperatures are likely to have risen, even if we manage to achieve net zero by the mid-century.

Worryingly, the warming of the planet also seems to be increasing the rate of emissions, particularly of methane, from natural sources such as drying peatlands. So, even if we are able to cut man-made CO_2 to zero, we may still see warming driven by natural processes that have been destabilised by the heating that has already occurred.

This means that we will need to find ways of capturing greenhouse gases from the atmosphere to compensate for the remaining emissions, or to try to slow the warming that is already happening. We can easily see circumstances in which we may wish to capture many billions of tonnes a year for these purposes.

Many people are not convinced this will be possible. Capturing greenhouse gases requires a lot of energy (which of course must be carbon free) in the form of heat and electricity. Usually, the sceptics don't question the technical feasibility of separating and then storing the CO_2 but accurately point to the tiny fraction of CO_2 in the atmosphere. They say that it is hugely inefficient in energy terms to separate out the 0.04 per cent of carbon dioxide from the air's oxygen and nitrogen.

Nevertheless, sucking CO_2 straight from air ('Direct Air Capture', or DAC in the jargon) is increasingly seen as attainable and – perhaps – financially possible. Whether or not we believe DAC is commercially feasible, the world needs urgently to push ahead with pilot plants that tell us whether the technology looks as though it can be scaled up rapidly and at reasonable cost.

The first companies commercialising DAC use very large fans to blow air across chemicals that naturally react with carbon dioxide. When these chemicals are fully loaded with CO_2, they are removed and heated. This reverses the reaction and the carbon dioxide is released into a sealed chamber. It is then taken away and stored, and the original chemicals can be put back to work capturing the next batch of CO_2.

IS DAC FINANCIALLY FEASIBLE?

Long dismissed as too expensive to be a plausible means of carbon sequestration, DAC is starting to attract serious attention. This is partly because of growing pessimism about the difficulties involved running conventional gas power plants efficiently if they have to be built to capture CO_2 from flue gases. But the increased interest is also for a positive reason. We are beginning to see convincing evidence that DAC may be much less expensive than previously thought.

The companies developing DAC plants are succeeding in reducing the energy needs of the capture process and are also improving the amount of heat needed to release the carbon dioxide from the chemicals with which it has reacted. And, as renewable electricity sources become cheaper, DAC is beginning to look more competitive.

The scientists behind Carbon Engineering, a Canadian company pioneering very-large-scale DAC, have published a paper suggesting that the cost of CO_2 capture may fall below $100 a tonne, once scaled up. The implications of this are very important: if the world (or a country) imposes a carbon tax of more than $100 per tonne of greenhouse gas emissions, then it will make financial sense for polluters to counterbalance their emissions with the capture of an equivalent amount of CO_2, for which they will get financial credit. If we find (as we may well) that avoiding greenhouse gas emissions completely is not possible, the world can hope that capturing CO_2 from the atmosphere will be available to help neutralise the remaining pollution.

THE MAIN CONTENDERS

Two start-up businesses are capturing most attention. Carbon Engineering is one. Backed by investment from major US oil companies and from Bill Gates, it is developing large-scale plants which both capture CO_2 and then use it, with hydrogen from electrolysis, to make synthetic fuels (as described in Chapter 1). It claims its first full-scale facility will capture half a million tonnes of CO_2 a year. (For comparison, a large coal-fired power station produces around 5 million tonnes of CO_2 a year.)

Somewhat controversially, Carbon Engineering is also cooperating with its shareholders in the fossil fuels industry to build a DAC plant that injects captured CO_2 into old oilfields. This will increase the amount of oil recovered from individual wells, apparently adding to the world's climate problems rather than reducing them. However, the companies claim that most of the captured CO_2 will stay permanently underground and will outweigh the emissions from the greater amount of oil that is extracted. I'm not convinced by this, but nevertheless think that Carbon Engineering was right to cooperate with the oil industry. The demand for CO_2 to enhance recovery of fossil fuels will provide an early market for the company and allow it to learn how to build its plants cheaply and quickly. It will also gain experience of using disused oil wells as key locations for storage of CO_2.

Carbon Engineering indicates that it can make a synthetic petrol substitute that costs less than US\$1 a litre if it builds a project of sufficient scale. At current oil prices, that is considerably more expensive than fossil fuel – which sold on wholesale markets at around 50 US cents at mid-2019 – but not much costlier than motor fuels made from agricultural products ('biofuels'), which are currently being extensively subsidised in Europe and the US.

Importantly, synthetic petrol will have far lower carbon emissions than biofuels if it uses renewable electricity as its energy source, and unlike biofuels it won't take up land that could otherwise be reforested or used for food production. The EU aims to have 10 per cent of motor fuels made from biological sources such as palm oil or maize by 2020 (at great environmental cost). Instead, it should consider providing backing to Carbon Engineering to build an experimental large plant.

Climeworks is the other promising start-up. Based in Zurich, this ambitious company is specialising in much smaller DAC installations than Carbon Engineering. Its first prototype plant piped the CO_2 it collected into a greenhouse. It is not unusual for greenhouses to use purchased CO_2 to improve yields, as plants grow bigger and faster with more in the air. Climeworks claims that its CO_2 has helped the greenhouse get 20 per cent more output of fruit and vegetables. The company is also building a plant in Switzerland that will provide all the CO_2 needed for a fizzy spring water made by Coca-Cola.

Even more interestingly, it has opened a site in Iceland that captures CO_2 and then sequesters it underground where the gas is chemically absorbed by basalt, a type of rock. The conversion of carbon dioxide into a calcite is permanent and without wider environmental impact. And basalt is extremely common around the world so CO_2 storage can be widely dispersed. The process is also relatively fast. When experiments began, it was unknown if CO_2 would take years or decades to be absorbed by basalt. It turned out to take just a few months.

Climeworks now offers a chance for air travellers to offset emissions by buying CO_2 sequestered in this way. Prices aren't cheap: a return flight from London to New York would come out at around \$1,100 for capturing and storing the tonne or so of CO_2. But if the company can scale out its technology, this price will go down sharply.

ROCK-SOLID CO2

The UK company Origen is hoping to commercialise its ideas for direct air capture employing a similar approach. Origen is using the natural tendency of certain minerals to react with CO_2 in the atmosphere much as Climeworks is using underground basalt. It uses the common mineral calcium carbonate, which we know as limestone, which it heats to make quicklime (calcium oxide). In the process, CO_2 is given off, which needs to be collected and either stored or used. When powdered limestone is exposed to air and water, it rapidly reabsorbs CO_2 from the atmosphere. Other researchers are looking at finely grinding up calcium- or magnesium-based rocks which are then spread out over the ground to absorb atmospheric CO_2.

The attraction of this route is that the weight of rocks available to capture CO_2 is almost unlimited and this process would be environmentally benign. There is little doubt the approach will work. Long ago in the earth's history, when CO_2 levels were sometimes far higher than today, the gas was slowly absorbed over millions of years by the weathering of rocks. All today's pioneers are trying to do is to speed up a natural process. However the research isn't yet advanced enough to give us any confidence about the likely cost of this approach compared to that of Carbon Engineering or Climeworks.

TAX PLUS DAC: A SOLUTION?

If we assume that one of the companies involved in direct air capture does get the cost of CO_2 down to $100 a tonne, and a carbon tax is applied at a similar level, DAC becomes effectively free. This means that companies trying to make low carbon synthetic fuels from hydrogen and carbon dioxide, such as Sunfire (see Chapter 1), will have a smaller bill for their materials and will be quicker to achieve parity with fossil fuels.

However, today's direct air capture industry is tiny, storing a few thousand tonnes a year, and we will almost certainly need to use the technology in future to sequester millions of tonnes. My guess is that, as the consequences of climate breakdown become obvious, we will need DAC both to produce synthetic fuels (which will then be burned, returning the pollutant to the atmosphere) and for sucking large amounts of CO_2 out of the atmosphere. The CO_2 will then need to be stored, in places such as in the depleted oil fields in the North Sea or by reacting it permanently with rocks.

We shouldn't underestimate the scale of the likely requirement to reduce CO_2 concentrations. At the moment, the amount of carbon dioxide in the atmosphere is rising about 2 parts per million per year. To capture the equivalent of just 1 part per million would require DAC to pull out over 2 billion tonnes of CO_2 a year. At a cost of \$100 a tonne, the total bill would be \$200 billion, or around 2 per cent of the global economy.

DAC enthusiasts suggest that the challenge of such new infrastructure is similar to the creation of sewage systems in the nineteenth century. These were initially resisted by many cities and governments but were swiftly recognised as a universal benefit and publicly funded. And, if the answer is to make small DAC units, such as Climeworks have developed, we already have several industries, such as air conditioning companies or motor car manufacturers that are entirely capable of producing these in great numbers.

All of which said, the better route for the world may be to swear off fossil fuels as quickly as possible, to protect and expand forests, and increase the carbon in our soils by improving our care for the land. DAC should be seen as a means of addressing the emissions we can't otherwise solve, but also as a faster way of getting to net zero.

CHAPTER 12

SHOULD WE GEOENGINEER?

Preparing to combat the worst consequences of climate change

What if we can't move to zero carbon fast enough, or the worst effects of climate breakdown arrive even more rapidly than we predicted? We need to plan and experiment now with means of adjusting global temperature – blocking the full energy of the sun through adjusting the transparency of the stratosphere. There are two main contenders for such geoengineering projects: cloud whitening, where spraying tiny particles of sea water above oceans causes more of the sun's energy to be reflected back into space; and 'solar radiation management' (SRM), where sulphur is released into the stratosphere, which has a similar effect to cloud whitening but much further away from the earth's surface.

These are both hugely controversial proposals, which will probably change rainfall patterns, and need urgent and stringent research. A very different and entirely benign solution is biochar, which is produced by burning plants or trees to high temperatures in the absence of oxygen. This carbon is then buried in a state of almost permanent storage; the biochar also improves soil fertility, and thus draws down more CO_2 via photosynthesis.

Nobody suggests that these techniques should be used as a substitute for reducing fossil fuel use. But the evidence is mounting that today's average warming of around 1 degree celsius has already caused unstoppable changes in ice cover, Arctic permafrost and the frequency of extreme weather. We probably need to simultaneously cut emissions and reduce the sensitivity of the planet's weather systems to the impact of higher levels of CO_2 stored in the atmosphere.

THE NEED FOR GEOENGINEERING

The world is already deeply damaged by climate change. Even at today's levels of warming, we observe far more severe effects than were predicted even a few years ago. This, combined with our failure to make more than a small dent in the rate of increase of greenhouse gases, should alert us to the need to use the full range of climate controls. We may need to change the planet so that it is less responsive to rising greenhouse gas emissions.

The list of potential 'geoengineering' techniques is long. Amongst other routes examined over recent years, scientists have looked at adding small amounts of iron to the southern oceans. Shortage of the element restricts today's growth of plankton. Greater numbers of plankton near the surface of the ocean would increase the capture of carbon, which, in theory, might then fall to the ocean floor when the organism died. Some scientists strongly advocate this technique, which would be very cheap. Others worry that it will severely disrupt oceanic ecosystems, perhaps by encouraging algal blooms.

Adding forest cover (see Chapter 9) is also a form of geoengineering because it pulls CO_2 from the atmosphere. Biochar, covered at the end of this chapter is an easy way to store the carbon extracted from air.

But, as noted, most attention is currently being paid to two approaches that do something very different. Both 'cloud whitening' and 'solar radiation management' reduce the amount of solar energy reaching the earth's surface. CO_2 levels in the atmosphere remain the same, but the extra heating caused by greenhouse gases is reduced. Neither proposal has yet been attempted in any significant way. Both deserve close examination.

CLOUD WHITENING

Cloud whitening works by spraying fine droplets of sea water to make thicker clouds over oceans. In normal conditions, clouds are less likely to form over the surface of seas because of the lack of tiny particles onto which water can condense. The geoengineering proposal is for small automatic and autonomous boats to criss-cross the seas spraying minuscule droplets that could enhance the size and the density of cloud cover.

One of the main proponents of this scheme, Stephen Salter of the University of Edinburgh, has calculated that just 10 cubic metres of tiny drops a second 'could undo all the [global warming] damage we've done to the world up until now'. He says it would take a fleet of 300 of his ships to wind back temperatures by about 1.5 degrees celsius. The scheme would cost hundreds of millions of pounds a year, but the price is tiny compared to the adverse impact of rising temperatures around the globe.

Why isn't the world trying out this scheme? The usual reasons crowd in − inertia, conservatism and the lack of immediate financial incentive. We also cannot be absolutely sure about the effects on agriculture and rainfall. Were, for example, Salter's boats to reduce the rains of the Indian monsoon, farmers there would suffer enormously. Nevertheless, small experiments with a small number of vessels would surely be worthwhile.

As with sulphur seeding (see below), we can hope that temperatures will be held down by these fine sprays. This will help reduce the pace at which ice melts and possibly restrain extreme weather. However, the droplets won't affect the level of CO_2 in the atmosphere, so problems of rising ocean acidification will not be solved.

SULPHUR AEROSOLS

Large volcanic eruptions throw up millions of tonnes of tiny particles, including sulphur compounds into the stratosphere many miles above the earth's surface. We have seen lower temperatures after such events around the world. The eruption of Mount Pinatubo in the Philippines in 1991 put about 16 million tonnes of dust high into the atmosphere and reduced average temperatures by about 0.6 degrees over the following year. The effect then dissipated as the particles eventually fell to earth.

We could try to achieve the same effect as a large volcanic eruption. That would mean lifting millions of tonnes of tiny particles of sulphur a year up to heights far greater than are reached by passenger jets. This would be a far more expensive operation than Salter's autonomous cloud-seeding boats. And it is probably more risky, because the sulphur particles may adversely affect the ozone layer. Again, we don't know the effect on global rainfall patterns, though a recent paper from Harvard University scientists suggests that only a tiny fraction of the world's surface would be likely to suffer adverse change. Once again, however, concerns over untoward effects are impeding progress towards even the small experiments that the Harvard group has suggested.

The UK can, and should, be a centre for research into geoengineering. It has the competences in chemistry and marine engineering to be the world leader.

BIOCHAR

This is a very different geoengineering proposal, looking to permanently store large volumes of carbon in the earth's soils in the form of biochar, a kind of charcoal made from a wide variety of plants, trees and other organic matter (such as animal dung).

In most of the plans being developed to build a global net zero economy, carbon capture and storage (CCS) plays a central role, particularly for the CO_2 coming from large power plants. The UK is no exception and looks to keep gas power stations working, storing the carbon dioxide in disused North Sea oilfields. More radically, many proposals see wood being used as a power station fuel in larger quantities than today, with similar plans for carbon capture. This route has the net effect of taking CO_2 out of the atmosphere because the carbon in wood originally came from the air.

The obvious concern is that CCS is still unlargely untried around the world and may be either expensive or impractical. We could, instead, make biochar from wood. Biochar is formed by heating dry plants or wood to very high temperatures in a simple furnace in the absence of air. The lack of oxygen means that the material doesn't combust, but energy-carrying gases are driven off from the wood and can later be burned to make electricity or heat. The only thing that remains in the biochar furnace is a charcoal-like substance, which is almost entirely pure carbon. Putting this into soil is an effective form of CCS, because biochar will usually last hundreds of years. Ploughing it into topsoil is simple.

The huge extra advantage is that in many types of soils the biochar substantially improves fertility. In carbon-poor soils in the tropics, yield improvements of 50 per cent or more are not uncommon. The reason is probably that, because biochar is filled with microscopic holes and has a huge surface area, it can provide a safe home to many beneficial soil organisms that aid plant growth. If we follow John Letts' recommendations (see Chapter 8) to grow grain without fertiliser in the UK, biochar could be extremely valuable because of its ability to hold many symbiotic fungi and bacteria, providing a real substitute for conventional fertilisers.

Arguments rage over the weight of carbon that could be stored each year in the form of biochar, but we are probably able to capture over 1 billion tonnes of CO_2 in this way, or several per cent of global emissions today. Whether biochar makes good financial sense for producers and farmers is still not certain, but a realistic carbon tax would almost certainly make it viable. What we urgently need is government-backed research to work out how to build a large industry that sequestered worthwhile amounts of carbon from the new woodlands proposed earlier.

RESERVATIONS ABOUT GEOENGINEERING

Many environmentalists are wholly opposed to geo-engineering. They point out the risks of unexpected effects, particularly from large experiments that affect the amount of solar radiation arriving at the surface, and note that scientists do not fully understand the climate system nor can predict what will happen. But, whatever our concerns about the possible disruption to rainfall, it needs to be pointed out that global heating is already hugely affecting where and how much it rains.

Professor Joanna Haigh, a leading climate scientist, says that geoengineering also introduces 'moral hazard'. If we rely on it to reduce the risks of climate breakdown, the strength of our commitment to cutting out fossil fuels may be weakened. Scientists like her tend to worry that we will use any excuse we can find to avoid the difficulties of transforming our societies so that our emissions fall to zero.

These are powerful points, but I still believe that the risks created by rising CO_2 levels are so great that we have no alternative to actively pursuing a range of geo-engineering options. As is repeatedly said, addressing the threat from climate breakdown requires us to use every tool at our disposal – not to spend precious time arguing about which techniques of carbon reduction are best.

CHAPTER 13

WHAT WE CAN DO OURSELVES

It's not just governments - our own actions can make a real difference

The proposals in this book cover action that needs to be taken to reduce all the UK's emissions of greenhouse gases close to net zero. If we want to do this by 2050, or before, we will need to act across all of these areas and probably use direct air capture to counterbalance the remaining emissions. Most of these are national actions that will need government planning. Some will need to be European or international. Perhaps, too, the world, working together, will also need to have put in place geoengineering plans to reduce the effect of the sun on the earth's atmosphere.

Generating political will to engage in such a wholesale transformation of the UK's economy will be intensely difficult, even with growing concern over climate breakdown. Without active support across society, it will be impossible. A New Deal for Climate will be needed to allow difficult, controversial and expensive measures that both reduce greenhouse gases and redistribute income. These proposals may seem too difficult to stand any chance of adoption, but net zero is probably unattainable without changes at least as radical as suggested here.

Technology alone will not take us to our goal, nor will changes in our patterns of consumption, nor a high tax on carbon. We need to do everything if we are to be successful in reaching zero carbon. Of course, my proposals may not prove to be the right ones, but those we do adopt are likely to be just as radical and equally disruptive to the way our economy currently operates.

The highest priority, because it addresses most of our current emissions, is to build an energy system that completely avoids fossil fuel use. This objective is perfectly attainable, but the cost in terms of investment may be as much as 4–5 per cent of national income every year for at least two decades. This would add at least 50 per cent to the total amount of spending on industrial capital – but at a time of near-zero interest rates the effect on the economy would be strongly positive. And, once the investment is made, energy will be much cheaper than ever before. The argument that acting on climate breakdown is 'too costly' is even less true than it has ever been.

Let's briefly think about the costs of the other measures mentioned in this book. Changes in diet will actually make families more prosperous, as they cut purchases of meat. Electric vehicles are currently more expensive, but this premium will be erased within less than five years. Car sharing will be far less costly than having a vehicle in every driveway.

Better insulation of houses will require significant investment, but the savings in fuel costs and improvements in health will match this cost. Steel and cement will almost certainly become more expensive. However, the impact on prices will be limited – using steel made with hydrogen will add about 1 per cent to the price of a car. Clothing, particularly in the UK, is currently cheap but this has had the unfortunate effect of making us wasteful and careless. Buying fewer items, but of far better quality and reusability, may save us money in the longer term.

CAN INDIVIDUALS MAKE A DIFFERENCE?

Much of this book is about policy changes that need to be made by government, both local and national, as well as large utilities companies. But individuals can also take action, both to put pressure on our governments and companies to change their policies, and to reduce our own carbon footprints. If we want the world to change, we need to demonstrate in our own lives that such changes are possible. 'Virtue signalling' it may be, but it affects the attitudes of others, and it makes us all think about how best to live a low carbon life. Witness Greta Thunberg sailing to New York – a small action in itself, but one that turned aviation emissions into an international issue.

Below, then, to conclude this book, are 20 suggestions for reducing an individual's carbon footprint in the UK. If we could act upon all of them, we might be able to reduce our responsibility for greenhouse gas emissions by 80 per cent or more – that is, to within a couple of tonnes of carbon neutrality. And, I would contend, we can do so without overwhelming changes (or 'sacrifices', as the press like to call them) to our Western lifestyle.

ENERGY

1. **Buy electricity from renewable sources.** Two types of retailers can sell you green power. The first owns and operates wind and solar farms. Their electricity tends to be expensive. Other retailers buy power from the wholesale market and issue certificates that guarantee the electricity is green. These companies offer cheaper tariffs. The average UK home heated by gas uses about 3,100 kilowatt hours of electricity a year. Look at your annual electricity and ask how your house compares. Switch now.

2. **Get your gas from anaerobic digestion.** Agricultural waste that is allowed to rot in the absence of air produces methane, which is the most important ingredient in natural gas. Green gas from the smaller energy retailers is expensive, but it saves carbon.

3. **Change your household lighting to LEDs.**
   This will almost certainly pay back the cost within a year, by reducing your electricity bills.

4. If you can afford it, install solar panels.

Now that subsidies have been abolished, payback will be slow, but your electricity generation will help green the grid. It probably doesn't make financial sense yet to install battery storage but this should change.

5. Buy really efficient washing machines and other goods when the time comes to replace your appliances. This helps reduce carbon even if you buy renewable electricity, because your lower power consumption means there is more green power available for others.

HOUSE INSULATION

6. Reduce heat loss around your home.

You probably won't save huge amounts of energy, but the cost of improving insulation around doors and windows, as well as blocking any air leaks between the floor and the skirting boards, is trivial. It all helps.

7. Seek advice on solid wall insulation and other more fundamental improvements. There may be subsidies available. Very-well-insulated homes should look at installing heat pumps as the means of keeping the house warm.

8. Keep temperatures at reasonable levels.

And turn down radiators in rooms you don't use. The average UK home heated with gas uses about 12,000 kilowatt hours. How does yours compare?

TRANSPORT

9. Fly less (and stay longer). Avoid flights whenever
you can. For those flights you have absolutely have to
take, contemplate buying an offset from Climeworks
(climeworks.com) that turns your emissions into
stone in Iceland. If that's too expensive, find a supplier
of carbon offsets that looks as though it will genuinely
plant more trees for your payment and maintain them.

10. Use public transport, or cycle or walk. whenever
possible. This will be good for you, and for carbon
emissions.

11. Drive an electric car (and share it if you can). If
you don't need a car on a daily basis, look at joining
an electric car-sharing club, or rent a car when
you need one. You'll be responsible for a much
smaller share of the emissions embedded in the car's
manufacture.

FOOD

12. Cut down on eating meat (or give it up entirely).
Particularly beef and lamb, which have the highest
carbon footprint.

13. Choose vegan alternatives to animal products,
as much as possible. There is a growing range of
alternatives, even for burgers. And they are healthier.

14. Avoid all airfreighted food and reduce purchases of
fresh foods that are out of season. If you can, grow
your own.

15. Buy less and avoid waste. This applies to food, goods, clothes and all other consumer goods.

16. Be fashion aware. Clothes have a high carbon footprint. Buy less, wear them more often, repair rather than throw away, shop second-hand and, when it's appropriate, look into renting.

17. Keep phones and other electronic goods as long as you can. Each year that you hold on to a computer might save hundreds of kilos of CO_2. If you need to upgrade computers or other goods, try to find them a new home. Most phones, for example, can be recycled.

18. Buy second-hand whenever you can, whether it's from Oxfam or other charity shops or on Depop, eBay or Freecycle. It makes a significant difference.

POLITICS

19. Join citizen activist groups such as Extinction Rebellion, Greenpeace and Friends of the Earth. Support climate change demonstrations and other forms of civic action.

20. Vote for politicians who prioritise action on climate change and commit to well thought-out and effective action, not people who just want to ban drinking straws. Write to your MP and councillors. Write to the Prime Minister and Chancellor. Make sure they know that people are concerned.

Above all, continue reading and talking about climate breakdown and ways we can work together to combat it. This is the most difficult challenge human civilisation has ever faced, and the future for humanity is dire unless many of us – not just a few scientists and activists – get involved in taking action. The solutions to the climate crisis are available, and the cost is bearable.

USEFUL BOOKS AND WEBSITES

A very brief selection

CLIMATE BREAKDOWN

Mark Lynas, *Six Degrees* (Harper Perennial, 2008). Lynas brought out this hugely influential book a decade ago – an exploration of what is likely to happen to the planet as it warms. In April 2020, an updated version is due to be published. It will be scientifically rigorous, beautifully written and probably very frightening.

OPTIONS FOR RENEWABLE ENERGY

David MacKay, *Sustainable Energy: Without the Hot Air* (Oxford UP, 2009). This classic book by the late David MacKay is now very much out of date, but the rigorous methodology makes it still worth reading.

Chris Goodall, *The Switch* (Profile Books, 2016). My own book on solar energy is three years old, so somewhat dated, but it contains useful detail on some of the topics in this book, including synthetic fuels.

Paul Hawken, *Drawdown* (Penguin 2018). This highly readable collection of expert essays outlines (and rates) 100 different ways of reducing emissions or collecting CO_2 from the air. It is a book that both inspires and informs.

The work of the UK's Committee on Climate Change is thorough and careful. I don't always agree with their conclusions, but the research is always robust and clear. Their website (www.theccc.org.uk) has papers covering most of the topics in this book.

HOUSING

The Energiesprong website (www.energiesprong.uk) is a useful resource for those interested in improving the UK's older housing stock.

FOOD

Mike Berners-Lee, *There is No Planet B: A Handbook for the Make or Break Years* (Cambridge UP, 2019). The best work on the food system and climate, including a substantial section on food. And anyone interested in the carbon footprints of modern lifestyles should also read Berners-Lee's *How Bad are Bananas* (Profile Books, new edition in Spring 2020).

CLOTHING AND STUFF

The website of Ellen MacArthur's foundation (www. ellenmacarthurfoundation.org) contains much of interest on the circular economy, and clothing in particular.

WRAP (www.wrap.org.uk), the waste minimisation institute, publishes consistently interesting material on cutting the use of resources.

The work of the **Energy Transitions Commission** (www.energy-transitions.org) is exceptional and includes detailed reports on sectors of the economy which are most difficult to decarbonise.

REFORESTATION

George Monbiot, *Feral* (Penguin, 2014). This is only in part about reforestation, alerting us more widely to the damage to the environment caused by sheep grazing and offering climate-friendly solutions for the restoration of our environment. It provides much of the background for discussion of woodland in this book.

GEOENGINEERING

Oliver Morton, *The Planet Remade: How Geoengineering Could Change the World* (Granta, 2016). The best analysis of how the world should think about geoengineering, as Morton dissects the feasible and unfeasible options.

CLIMATE CHANGE MITIGATION

I hugely admire the work of **Carbon Brief** (www.carbonbrief.org), a website that publishes detailed and insightful analysis of what is going on in the UK and elsewhere in the fields of renewable energy and the science of climate change.

Anyone interested in keeping up with new research should also watch **Carbon Tracker** (www.carbontracker.org), which reports on stranded assets; **Energy and Climate Intelligence Unit** (www.eciu.net), which covers energy and climate change news; and, from Australia, **RenewEconomy** (www.reneweconomy.com.au), an exceptional resource for keeping up with events in perhaps the world's most advanced renewable energy market (despite its pathological continued emphasis on coal).

My own website, **Carbon Commentary** (www. carboncommentary.com), also has useful updates on climate issues, along with source notes for this book.

GREEN NEW DEALS

Ann Pettifor, *The Case for the Green New Deal* (Verso, 2019). The UK-based economist looks at how countries can best allocate capital towards projects that improve low carbon infrastructure. Jonathan Ford in the *Financial Times* summarised Pettifor's conclusions by saying that 'she sees the nation state as eminently capable of financing decarbonisation'.

Naomi Klein, *On Fire: The Burning Case for a Green New Deal* (Allen Lane, 2019). Klein brings a transatlantic perspective and a more radical emphasis on changing the very structure of our economies. She argues that conventional capitalism is incapable of dealing with the threat from climate breakdown.

In this book I have tried to argue for a more moderate stance, not because of any faith in the ethical standards of corporations, but rather a conviction that rapid growth of zero carbon alternatives depends on the managerial and innovation skills of conventional large companies.

THANKS

The idea for this book came from Mark Ellingham at Profile Books. This is the fourth title I have published with him and his colleagues. As always, he has been an enthusiastic and imaginative editor. Many of the best sentences were actually written by him.

Thanks also to the team at Profile Books: cover designer Peter Dyer, text designer Henry Iles, proofreader Nikky Twyman, indexer Bill Johncocks and publicist Drew Jerrison.

I'm especially grateful to all those who responded to my requests for interviews and I hope their names are all mentioned in the text. Mike Berners-Lee was very generous in carefully reading and correcting the draft text. Professor Malcolm McCulloch was very generous with his time talking to me about the many new local energy initiatives within Oxfordshire. Remaining errors are all my responsibility, of course.

This book is dedicated to my wife Charlotte Brewer, a constant supporter, and to our daughters Alice, Miriam and Ursula.

INDEX